Contents

Preface

The title 'BS 449' is recognized throughout the world as the main British code of practice devoted to the design of structural steelwork. First published as a byproduct of the activities of the Steel Structures Research Committee in 1931, BS 449: *The Structural Use of Steel in Buildings* has been revised, extended and amended to take account of improved understanding of structural behaviour, changes in fabrication techniques and the requirements of new forms of construction on several occasions. The most recent metric edition is dated 1969. Some two years prior to this a decision was taken to begin work on a completely revised version which would not only update the document's detailed design procedures but would recast these into the more progressive limit states format. Of course such a move was not universally well received by structural designers; it is still unpopular with a section of the profession today. It did, however, represent a course of action that either has been or is being pursued by most of the main UK structural codes as well as by steelwork codes in many other parts of the world.

The author first became directly involved in the production of this new code in 1971. It was through this contact that the idea for a textbook explaining the material of the code to both students and practising engineers gradually developed. Work on the text began at about the time that the original draft for comment – the so-called B/20 Draft – was issued in 1977. Because the reaction to B/20 was sufficient to require substantial alterations and re-drafting, the appearance of the code with its new designation BS 5950 has taken several years. Completion of the text has, of course, had to await finalization of this document.

The book is not intended to be a commentary upon the new code; that document has been prepared by Constrado as part of their role in producing the actual text for both the code and the supporting material. Rather, it is a textbook on structural steel design according to the principles and procedures of BS 5950. As such it is aimed principally at students of steelwork design – whether they be undertaking courses in universities or polytechnics or, having successfully completed this phase of their career, are working in practice and want to update their knowledge. Therefore it is hoped that the material will be both self-contained and suitable for private study. It does, of course, make frequent reference to particular clauses in the code itself.

In writing this book the author has benefitted enormously from various forms of interaction with a large number of organizations and individuals. These have included those responsible for the preparation and drafting of the code, teachers of steelwork design, representatives of overseas steelwork code committees, engineers in practice and delegates on various post-experience courses and seminars on either the new code or on steelwork design in general. Frequently, seemingly small points raised in discussion have provided the impetus for a change in the text or for the inclusion of an additional point of explanation. To all of these the author is most sincerely grateful.

The manuscript was prepared using the facilities of the Department of Civil and Structural Enginering in the University of Sheffield. The text was typed by Miss Janet Stacey whose patience in dealing with the numerous revisions is greatly appreciated.

D.A. Nethercot

Notation

A	cross-sectional area
A_e	effective area of a tension member
A_g	gross area of section
A_n	net area of section
A_s	shear area of bolt
A_t	tensile area of a bolt
a	throat size of fillet weld, projection of baseplate, depth of haunch, distance between member axis and restraint axis, spacing of vertical web stiffeners
a_1	net area of connected leg of section
a_2	gross area of unconnected leg of section
B	width of section
b	clear width of plate
b_e	effective width of slender plate
b_1	length of stiff bearing
D	overall depth of section, hole diameter
d	diameter of bolt, clear web depth
E	Young's modulus
E_{st}	strain hardening modulus
e	distance between centroid and extreme fibre of a section, end distance
e_x, e_y	eccentricity of axial load
F	axial load
F_q	difference between actual shear in web adjacent to stiffener and shear capacity of web
F_s	applied shear
F_t	applied tension
f	factor to allow for bending effects
f_a	axial stress
f_b	bending stress
g	gauge of holes
h	storey height
h_1	height to eaves for a portal frame
h_2	height from eaves to apex for a portal frame
I	second moment of area
I_s	second moment of area of web stiffener
I_y	second moment of area about y-axis
I_x	second moment of area about x-axis
K_e	factor used to define A_e, see eqn 3.1
K_f	measure of flange contribution

K_s	factor on slip resistance of HSFG bolt, see eqn 7.5
k	effective length factor
k_1, k_2	restraint parameters for column in a continuous frame
L	length
L_m	maximum unbraced length for a member in a plastically designed structure
L_t	maximum distance between torsional restraints in a plastically designed structure, see eqn 8.5
ℓ	effective column length
$\overline{M}$	equivalent uniform moment
M_{ax}	buckling resistance moment for combined axial load F and major-axis moment M_x
M_{ay}	buckling resistance moment for combined axial load F and minor-axis moment M_y
M_b	buckling resistance moment
M_c	moment capacity of section
M_E	elastic critical moment for lateral–torsional buckling
M_f	moment capacity of girder flanges
M_p	fully plastic moment of cross-section
M_{rx}	reduced moment capacity about x-axis in the presence of axial load F
M_{ry}	reduced moment capacity about y-axis in the presence of axial load F
m	equivalent uniform moment factor
n	slenderness correction factor, ratio F/Ap_y
n_1	length under point load due to load dispersion, see eqn 5.13
P	axial load
P_c	compressive capacity
P_{cr}	elastic critical load
P_{cx}	axial capacity for xx buckling
P_{cy}	axial capacity for yy buckling
P_0	maximum shank tension for bolt
P_s	shear capacity of a bolt, slip resistance of an HSFG bolt
P_t	tensile capacity of a bolt
P_w	web buckling capacity
P_y	squash load of column
P_z	capacity of vertical web stiffeners
p_a	axial stress
p_b	bending strength
p_{bs}	bearing strength of connected parts
p_{bc}	bearing strength of bolt in plate
p_{bx}	maximum bending stress due to M_x
p_{by}	maximum bending stress due to M_y
p_c	compressive strength
p_s	shear strength of bolt material
p_y	design strength of steel
p_w	strength of fillet weld
Q	load effects
q_b	basic (tension field) shear strength
q_{cr}	critical shear strength
q_e	elastic critical stress for shear buckling
q_f	flange dependent shear strength
R	structural strengths (resistances)
r_{aa}	radius of gyration of an angle about an axis through the centroid parallel to the gusset
r_{min}	minimum radius of gyration
r_{vv}	radius of gyration about minor principal axis
r_x	radius of gyration about x-axis

r_y	radius of gryration about y-axis
S_{rx}	reduced plastic section modulus about x-axis in the presence of axial load F
S_{ry}	reduced plastic section modulus about y-axis in the presence of axial load F
S_x	plastic section modulus about x-axis
s_p	staggered pitch of holes
T	flange thickness
t	plate thickness, web thickness
U_s	specified minimum ultimate tensile strength of steel
u	buckling parameter
V_b	web capacity allowing for 'flange dependent contribution' (tension field)
V_{cr}	shear capacity of web based on shear buckling
V_{tf}	contribution to web capacity due to tension field action
V_{ult}	ultimate capacity of web based on tension field action
v	lateral deflection, slenderness factor
W	transverse load
W_p	value of W at plastic collapse
W_y	value of W at initial yield
w	pressure on underside of baseplate, distributed load on beam
x	torsional index
Y_s	specified minimum yield strength of steel
Z	elastic section modulus
z_1, z_2	index, see eqn 6.4
α	$2y_c/d$, measure of bending present in a compressed plate, see Table 8.1
β	M_1/M_2, ratio of end moments $(-1 \leqslant \beta \leqslant 1)$
$\beta_1, \beta_2, \beta_3$	limits on plate slenderness for plastic, compact and semi-compact cross-sectional behaviour respectively
γ_e	factor of safety in permissible stress design
γ_f	partial factor on loading
γ_m	partial factor on materials
γ_p	global load factor in plastic design, partial factor on structural performance
δ	deflection
ϵ	strain
ϵ_{sh}	strain at onset of strain hardening
ϵ_y	strain at initial yield
λ	slenderness
λ_c	ℓ/r_{min} for main member of compound strut
λ_{cr}	load factor for elastic instability of frame
λ_{TB}	effective lateral–torsional slenderness for a member, allowing for tension flange restraint, see eqn 8.2
λ_{Tc}	effective minor axis slenderness for a member, allowing for tension flange restraint, see eqn 8.2.
λ_{LO}	value of λ_{LT} below which $M_b = M_p$
λ_{LT}	$\sqrt{\dfrac{\pi^2 E}{p_y}}\ \sqrt{\dfrac{M_p}{M_E}}$, lateral–torsional slenderness
λ_w	$(0.6p_y/q_e)^{\frac{1}{2}}$, web slenderness
μ	slip factor
σ_p	stress at limit of proportionality
σ_{ult}	ultimate tensile stress
σ_y	yield stress
σ_{yL}	lower yield stress
σ_{yu}	upper yield point

Index of references to clauses in BS 5950

1
Steel as a structural material

Steel is the most widely used structural metal. Its popularity may be attributed to the combined effects of several factors, the most important of which are: it possesses great strength, it exhibits good ductility, it has high stiffness, fabrication is easy and it is relatively cheap. Good examples of structural steelwork design seek to exploit each of these features to the full.

Steel's high strength permits heavy loads to be carried by relatively small members thereby reducing the self-weight of the structure. This reduction in dead load facilitates the construction of the large clear spans needed, for example, in sports halls. At points of very high stress such as in the immediate vicinity of a bolt, yielding of the material will enable the load to be redistributed smoothly and safely; this process makes use of the property known as ductility. All structures will deform to a certain extent when loaded – even when such loading consists only of the structure's own self-weight. Because steel possesses great stiffness (as measured by its modulus of elasticity E) these deflections will not normally be large enough to require special consideration. Steel may be worked in the fabricating shop in a number of ways, for example sawing, drilling and flame cutting; it may also be joined together by welding. Finally the price of steel is substantially less than that of any possible competing metal; for instance aluminium costs about three times as much as the basic structural grades of steel.

In the civil engineering field steel is in competition principally with reinforced and prestressed concrete, timber and brickwork, with many designers seeing the usual choice as being simply between steelwork and concrete. The reasons most often given for selecting a steel structure are listed in Table 1.1. Since the requirements of individual projects may vary so much it is not possible to provide simple rules for selecting the 'best' solution, even on the limited basis of initial cost. Rather, the designer must consider each of the factors present, must decide on their relative importance and must then use his judgement and experience to decide upon the most appropriate solution.

1

Table 1.1
Advantages of steel structures

Item	Comments
Ease of erection	No formwork
	Minimum cranage
Speed of erection	Much of the structure can be prefabricated away from the site
	Largely self supporting during erection
Modifications at a later date	Extensions/strengthening relatively straightforward
Low self weight	Permits large clear spans
Good dimensional control	Prefabrication in the shop ensures accurate work

As shown in Table 1.1, steelwork will often be the choice when questions of ease and speed of erection, for example a bridge over a busy railway line that can only be obstructed for short periods, large clear spans such as a grandstand for which no interference with visibility can be tolerated, or subsequent modifications to say a workshop which may be extended in size or into which additional cranage may be installed, are important. However, other factors, namely the non-availability of suitable aggregate locally or special architectural features may also control the decision.

1.1 PRODUCTION

Steel is made by refining iron which has itself been smelted from the basic iron ore in the blastfurnace. Ironmaking has changed little in principle in over 2000 years, although the actual techniques employed as well as the scale of production have, of course, altered considerably. Nowadays blastfurnaces operate continuously over a period of several years, producing up to 8000 tonnes of molten iron every 24 hours [1]. Iron ore, coke, limestone and sinter (a mixture of ore, coke and limestone that has previously been roasted together to remove some of the volatile matter) are fed into the top of the furnace. Air is blown through to increase the temperature, the oxygen content reacting with the hot carbon in the coke to form carbon monoxide which in turn releases the iron. The molten metal is periodically tapped off from near the base for subsequent use as the basic raw material employed in steelmaking.

1.1.1 Steelmaking

Three main processes exist for the production of steel. The oldest of these is the open hearth process [2]. Because it is slow and therefore relatively uneconomic it is steadily being replaced by the basic oxygen (BOS) process [3] and the electric arc method [4], which was originally devised to produce high-quality steels requiring precise control over their composition. Today most structural steel is made using the BOS process shown in Fig. 1.1. This commences with the operation known as charging, in which a mixture of molten iron and up to 30% scrap is poured into the top of the BOS vessel. High-purity oxygen is then blown in at great speed using a water-cooled lance. This combines with excess carbon and other unwanted impurities which then float off as slag.

During this time the temperature and chemical composition are carefully monitored and when both are adjudged correct the steel is tapped into a ladle. At this stage a sample is taken for chemical analysis and subsequent examination of its physical properties; the results of these appear on the mill certificate which must be provided to the eventual purchaser of the steel. From the ladle the still molten metal is cast into moulds where it solidifies into the ingots which will be taken to the mill to be rolled into plates, structural sections, bars and strip. This takes about 40 minutes (compared with 10 hours in the open-hearth method) and may involve an initial charge of more than 350 tonnes [5].

1.1.2 Rolling

At first sight it may appear strange that molten steel should first be cast into ingots which must then be reworked into usable shapes, rather than be cast immediately as plate, bar, etc. It is, of course, precisely this variety of pro-

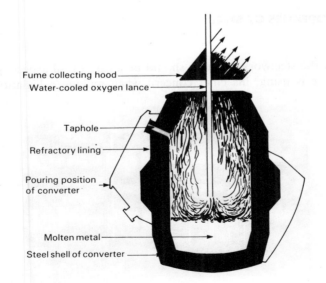

Fig. 1.1 Basic oxygen steelmaking (BOS) process. (*After ref. 5.*)

ducts, as well as the practical difficulties associated with the casting of shapes such as wide thin sheets, that dictates the need for other processes. Moreover, the reheating, together with the actual mechanical working received during rolling, modifies the steel in such a way that its tensile strength is considerably enhanced. The most common treatment is hot rolling in which the steel is squeezed between a pair of rotating cylinders termed rolls. In this way the original ingots, weighing anything between 5 and 40 tonnes, are reduced in stages down to plate, strip (thin plate), sections, bars, wire, etc.

The sequence of operations involved in hot rolling [6] commences with the ingot being heated in a soaking pit for between two and eight hours. This is necessary in order to ensure that it attains an even temperature throughout (even when it 'solidifies' in the mould its size is such that the centre will still be molten). From here the ingots proceed to the primary rolling phase in which they are passed repeatedly through heavy rolls of the type shown in Fig. 1.2. Each pass, of which there may be up to fifteen, reduces the thickness by as much as 50 mm. On emerging, the long slab or bloom has its ends cropped before passing through a second stage of rolling in the billet mill. The steel leaves here in the form of 10 m lengths of semi-finished material which are then inspected both visually and ultrasonically for surface and internal defects, such as cracks, blow-holes and major slag inclusions. It is then reheated by passing through a series of furnaces until it reaches the final series of profiled rolls – so-called because, as shown in Fig. 1.3, they operate on all four edges – which turn it into recognizable structural shapes. Final shaping of flat products (plate, sheet and strip) usually takes place in a four-high mill, in which the presence of the outer rolls reduces bending of the working rolls.

1.2 PROPERTIES OF STEEL

Although the steelwork designer should be aware of all aspects [7] of the material he is using, his chief concern when making calculations which

Fig. 1.2 Rolls used for primary rolling of steel ingots. (*After ref. 6.*)

Fig. 1.3 Model of profiled rolls used in final rolling of H-sections. (*After ref. 6.*)

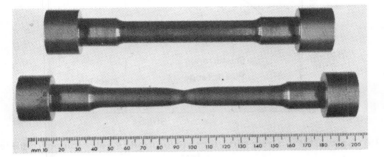

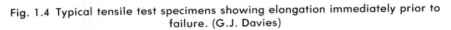

Fig. 1.4 Typical tensile test specimens showing elongation immediately prior to failure. (G.J. Davies)

attempt to assess the load-carrying capacity of a particular member will normally be material strength. This property is usually measured in a tensile test in which a small coupon of material is pulled in a testing machine until it fractures. Such tests also furnish useful information on material stiffness and ductility. Guidance on the tensile testing of structural steel specimens is given in BS 18, which covers items such as specimen dimensions, testing speed and the proper interpretation of the results. Figure 1.4 shows some typical tensile test specimens, while Fig. 1.5 gives details of their recommended proportions.

The results of a tensile test are normally quoted in terms of a stress–strain curve for the material. A typical curve for structural mild steel is shown in Fig. 1.6 with an enlarged version of the most important, initial portion being given in Fig. 1.7.

When load is first applied the specimen responds initially in a linear elastic fashion and obeys Hooke's law. Stress is directly proportional to strain and removal of the load results in the strain falling to zero. The slope of this straight-line portion is Young's modulus, E. As the strain is increased a point is reached at which the curve tends to depart from linearity. The stress at which this occurs is termed the 'limit of proportionality' σ_p and its presence is often difficult to detect. Further straining will result in the steel yielding at a

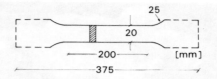

Fig. 1.5 Typical dimensions of a rectangular cross-section tensile test specimen according to BS 18: Part 2: 1971.

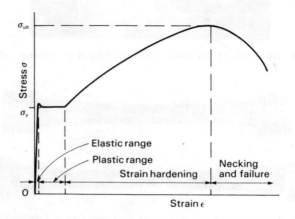

Fig. 1.6 Typical stress–strain curve for structural mild steel obtained from a tensile test.

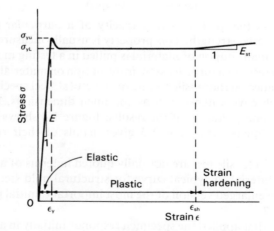

Fig. 1.7 Enlarged, slightly idealized initial portion of the tensile stress–strain relationship for structural steel.

stress equal to the upper yield point σ_{yu}. Once this stage has been reached the material no longer behaves elastically; even complete removal of the load will leave some permanent deformation in the specimen. As the strain proceeds beyond σ_y so the stress drops slightly to the lower yield stress σ_{yL} (often called simply yield stress σ_y). The margin between σ_{yu} and σ_{yL} depends on the type of steel and also on the speed at which the test is conducted, with a typical value of the ratio σ_{yu}/σ_{yL} for normal structural steel being about 1.05–1.10, although higher values have been observed in tests involving particularly low rates of straining [8].

Tests at too high a rate may well result in a complete failure to observe an upper yield point [9]. Once the stress has dropped to σ_{yL} it remains sensibly constant for considerable increases of strain as shown in Fig. 1.6. During this phase plastic flow of the material is taking place, the extent of which is a measure of the ductility of the material. Typical structural steels possess yield plateaus of at least ten times the strain at yield. Eventually yielding ceases and the stress starts to rise as the material strain hardens. The initial slope of this part of the curve is termed the strain hardening modulus E_{st}. It is much less steep than the elastic part, with E_{st}/E being typically between about 1/30 and 1/100 [8]. Eventually a maximum is reached on the stress axis; this corresponds to the ultimate tensile stress, σ_{ult}. Thereafter stress appears to decrease until fracture finally occurs. However, the real stress is actually still increasing; an apparent decrease is seen because the plotted quantity (often termed the 'engineering stress') is calculated using the original area whereas once σ_{ult} has been attained the actual area of the specimen reduces quite rapidly, a phenomenon known as 'necking'.

Although it is possible to conduct compressive tests on coupons this is complicated by the need to prevent the specimen buckling sideways. The results of such tests show the behaviour of most structural steels to be very similar in compression and tension, with the compressive yield stress being some 5% higher on average than the tensile value [10].

Ductility is measured by the percentage elongation, i.e. the increase in length divided by the original length measured over a standard gauge length (50 mm or 200 mm) obtained in the above test. Values as high as 20% of the original specimen length may be obtained. It is this property that enables small regions that are very highly stressed to yield, thereby relieving this concentration of stress without undue distress to the structure as a whole. Adequate ductility is also a pre-requisite for the use of the plastic design methods described in Chapter 8.

1.2.1 Comments on yield stress

Reported values of the mechanical properties of the structural grades of steel used in the UK are listed in Tables 17 (plates), 19 (sections) and 21 (hollow sections) of BS 4360. Compliance with these is normally the responsibility of the steel producer, who will seek to ensure that this has been achieved through tests on samples taken from each batch of steel. The results of these tests are shown on the mill certificate. In cases where such tests reveal a shortcoming it is possible that the batch may be sold as a lower-quality grade, providing, of course, that it meets the minimum standards for that grade. Because of this it is sometimes possible for the user to find that in several respects his material possesses better properties than he expected. While this may be of significance to the researcher (high-strength material means that he will require higher loads for his laboratory tests) it should not worry the designer; indeed because designers normally use specified properties, only rarely calling for their own materials tests in the case of steel, it is something of which he will probably remain unaware.

Tensile tests performed by the manufacturer are frequently referred to as 'mill tests'. It is usual for them to be conducted at a fairly high rate of loading (a strain rate of 0.0025/s is mentioned in BS 18). This is important because the yield stress of steel is strain-rate dependent [9], namely the results obtained from a tensile test will be influenced by the speed at which that test is conducted. Figure 1.8 illustrates this point.

By stopping the separation of the jaws of the testing machine so that the specimen is in effect being loaded at zero strain rate for a short period, it is possible to determine a minimum value of lower yield stress a few per cent below that which corresponds to straining on the yield plateau at a finite rate. This is termed the static yield stress and is illustrated in Fig. 1.9. Because the majority of loads on civil engineering structures are usually considered to be of an essentially static nature, it is generally accepted that the static yield stress is the most appropriate basis for normal design calculations. Mill tests, however, tend to measure a higher, dynamic figure which, because of the procedure employed, will also contain some upper yield point effects [8]. It is therefore comforting to find that the average values of yield stress obtained from mill test results may be expected to lie significantly above the guaranteed minimum values of BS 4360. As an example, Fig. 1.10 shows the results of tests on American ASTM A7 steel (broadly equivalent to Gr. 43) plotted as a frequency distribution. From the interpretation of these data given by McGuire [11] it would appear that only about 1% of mill test results

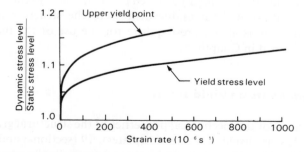

Fig. 1.8 Effect of strain rate on upper yield point and yield stress of structural steel. (*After ref. 8.*)

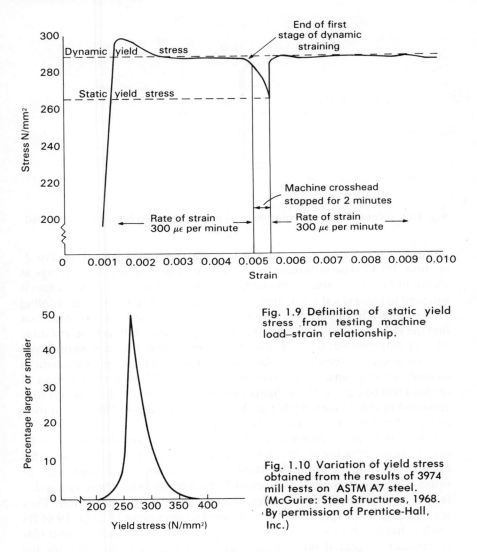

Fig. 1.9 Definition of static yield stress from testing machine load–strain relationship.

Fig. 1.10 Variation of yield stress obtained from the results of 3974 mill tests on ASTM A7 steel. (McGuire: Steel Structures, 1968. By permission of Prentice-Hall, Inc.)

fall below the specification value of 226 N/mm². However, if it is accepted that the static yield stress lies some 15% below the typical mill test value [12] then a mill test result of 260 N/mm² is necessary to ensure a static value of 226 N/mm². From Fig. 1.10 it may be seen that some 40% of samples fall below this figure. However, since the majority of these are not more than about 10% low, the net effect when averaged over a complete design is not likely to prove significantly. Both the 15% difference between mill test results and the static yield stress and the shape of the distribution shown in Fig. 1.10 have been confirmed by a Swedish study [8].

In the case of structural sections the difference between material strength as indicated by the mill test and the real, that is static, yield stress may also be

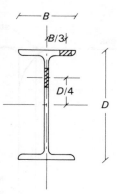

Fig. 1.11 Location of tensile specimens in a steel I-section according to BS 4360.

influenced by the position from which the specimens are taken. For I-sections, BS 4360 permits these to be cut from either the web or the flange as shown in Fig. 1.11. Since web material is thinner than that of the flanges it will tend to possess a slightly finer grain structure as a result of faster cooling after rolling. The importance of this to the structural designer lies in the fact that the yield stress will be higher [8, 12]. For a series of UB sections in Grade 43 steel differences of up to 16% of flange yield strength have been observed [13]. In most situations it is the flanges of an I-section that contribute most to its load-carrying capacity. For instance, even when it is used as a tie, most of the load will be carried by the flanges simply because most of the area is concentrated in the flanges, while members in bending derive virtually all their strength from the contribution of the flanges.

It is clear from the above discussion that the structural designer must be careful in selecting the appropriate value of material strength for use in his calculations. For the usual structural steels according to BS 4360, a set of values of design strength p_y is given in *Table 6* of BS 5950. These values differentiate between different grades, thicknesses and types of section. They are based upon the specified minimum yield stresses given in Table 19 of BS 4360, suitably adjusted by a partial safety factor on material strength (for explanation of partial safety factors see Section 2) so as to allow for the effects of all of the factors discussed above.

1.2.2 Residual stresses

Figure 1.5 shows that the mechanical strain at yield for structural steel is of the order of 1%; this is approximately the same as the expansion produced by placing a piece of steel in boiling water, i.e. increasing its temperature to 100°C. Much higher temperatures, typically 600–700°C, are involved in the rolling of steel, while members fabricated by welding (possibly using material that has previously been flame cut) will be subject to a further application of heat. Moreover, this heating will be applied locally to selected parts of the cross-section. Cooling of the heated material will always take place unevenly, even for the hot-rolled member placed on the cooling bed after rolling, for which air will reach the extremities, such as the flange tips of an I-section, more readily.

The net result of these processes of uneven heating and cooling is that structural members will normally contain residual stresses. Although these may be removed by subsequent reheating and slow cooling the process is expensive and is limited to special components like pressure vessels, for which the presence of residual stresses is known to be particularly unwelcome. As a general rule those parts of the section which cool first will be left in residual compression, while those that cool more slowly will contain residual tensile stresses. The region adjacent to a weld will normally be stressed up to yield in tension with balancing residual compression elsewhere in the section. Figure

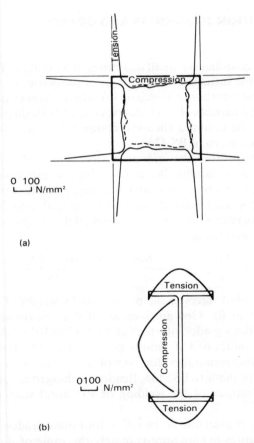

Fig. 1.12 Typical measured patterns of residual stress in structural sections. (a) 400 × 400 × 5 mm corner welded box section [14]. (b) 250 × 146 UB rolled I-section [15].

1.12 illustrates typical patterns of residual stress for a rolled I-section and a welded box.

Because residual stresses must themselves be in equilibrium, their effect on structural behaviour is limited; the most important consequence for statically loaded structures is to cause the member to behave as if it possesses a non-uniform distribution of yield stress over its cross-section. This is particularly important for compression members, for which those regions containing residual compression yield at loads producing an applied stress of less than σ_y. Members in bending also yield early and therefore tend to deflect more [16]. The presence of residual stresses also tends to lessen a member's resistance to the growth of cracks, whether this occurs in a stable manner due to the action of fluctuating loads (fatigue), or in a unstable fashion by the process known as brittle fracture [11].

1.3 COMPOSITION TOUGHNESS AND GRADES

Structural steels contain very small quantities of a number of elements, each of which has some influence on the physical properties of the steel. Most important of these is carbon; an increase in carbon content causes increases in both strength and hardness but at the expense of both ductility and toughness. Details of the required chemical compositions for all UK grades of structural steel are given in BS 4360.

Chemical composition also affects a steel's suitability for welding, a property known as weldability. Because welding is so extensively employed in the fabrication of structural steelwork it is important that the steel used be capable of being welded without the need for special, and therefore expensive, welding procedures. A measure of weldability is the so-called carbon equivalent, C.E. defined as

$$\text{C.E.} = \text{C} + \frac{\text{Mn}}{6} + \frac{\text{Cr} + \text{Mo} + \text{V}}{5} + \frac{\text{Ni} + \text{Cu}}{15}$$

in which each symbol refers to the proportion by weight of that particular element. Table 3 of BS 4360 gives values of C.E. of between 0.39% and 0.54% for the various grades, of which about 0.15–0.30% will be due directly to carbon. Low values of C.E. imply good weldability. For details of the appropriate welding techniques, for example metal arc, submerged arc, etc., reference should be made to BS 5135. Readers wishing to acquaint themselves with the basic features of the welding of structural steel should consult reference [17].

Structural steel is available in the UK in four main grades: 40, 43, 50 and 55, where the figures denote approximately the value of σ_{ult} in kg f/mm². Each grade is subdivided in descending order of C.E. values from A to E. The main structural grade is 43 (mild steel) with perhaps a 10% use of Grade 50, mostly in bridgework. Grade 40 is rarely used·as is the highest strength Grade 55 for which σ_y is of the order of 430 N/mm², approaching twice that of Grade 43. Present pricing policy is such that Grade 50 costs about 10–15% more than Grade 43 with the price of Grade 55 being a further 25% higher.

Toughness is necessary in structural steel in order to avoid the phenomenon known as brittle fracture. This can cause complete failure by the very fast propagation of a small crack, often in regions of comparatively low stress. Much has been written about brittle fracture since the first failure was identified in 1886 [11]. An account of some of the subsequent failures is given by McGuire [11], who also describes the metallurgical processes involved. From the designer's point of view the most satisfactory way of dealing with brittle fracture is to reduce the likelihood of its occurrence by a sensible choice of material. Providing the structure will not be subject to combinations of situations which are conducive to brittle fracture, such as low temperature, thick plates with mutually perpendicular welds stressed in the through-thickness direction and fast rates of loading, then this is not too difficult.

The approach taken by BS 5950 is that brittle fracture is unlikely for routine applications of structural steelwork in the United Kingdom. Thus *Cl. 2.4.4* directs the reader to a table of 'safe' maximum material thicknesses, noting that some modification will be necessary if certain circumstances prevail, including welded fabrication and stresses in excess of 100 N/mm² in material of more than one half the limiting thickness. For potentially critical situations such as welding details which induce a high degree of restraint, the designer is advised to seek specialist guidance. Since the method by which this may be obtained is not specified, the onus rests with the designer to use his judgement and experience backed up by the advice of a materials specialist if the circumstances are thought to warrant it [18].

For structural sections Grade 43A material may be used in thicknesses up to 25 mm for internal applications (assumed minimum temperature of −5°C), or 15 mm for external applications (assumed minimum of −15°C), beyond which Grades 43B(30 mm and 20 mm) or 43C (50 mm and 40 mm), 43D or 43E (both 50 mm) are required. Several of the limits in *Table 4* are simply the maximum thicknesses for which toughness data exist; thicker material may be used if it can be shown to possess adequate toughness. However, only four of the largest UC sections have flanges which are more than 50 mm thick [19]. The basis of *Table 4* is that the steel exhibits sufficient energy absorption when subject to a Charpy vee-notch impact test [18]. This is a standard material test in which small bars containing a notch are fractured by a heavy pendulum, the energy required being determined from the swing of the pendulum. Results are normally quoted as Charpy values C_v. Since they are currently affected by temperature, C_v values must be related to the testing temperature and a figure of −5°C is often taken as representing the minimum service temperature. More detailed information on the significance of Charpy test values and their relationship with true fracture toughness, as indicated by the application of the recently developed technique of fracture mechanics, is given by Burdekin [20] in a paper describing the basis for the toughness requirements for bridge steel in the UK.

1.4   FATIGUE

In structures subject to a very large number of cycles of fluctuating load, typically at least 100 000 load applications, failure may occur by the continued growth of cracks in the material at stresses well below those necessary to cause ordinary static yielding or collapse. Such behaviour is termed fatigue. Most civil engineering structures do not experience loads approaching their design load very frequently. Of course, there are certain exceptions, in crane girders, railway bridges, and offshore structures subject to wave loading. However, even ordinary wind loading does not normally provide sufficient repetitions unless the structure is susceptible to wind-induced oscillations. When it is realized that 100 000 cycles corresponds to ten applications a day for more than 25 years, it becomes clear that fatigue is unlikely to be a problem for ordinary building structures. It is more significant for bridges, although even here, since fatigue is largely dependent upon stress range, i.e. the difference between the maximum and minimum stresses experienced in service, many bridges will not receive sufficient applications of load heavy enough to produce the necessary large changes in stresses.

For the design of crane supporting structures BS 5950 refers the engineer to BS 2573; for more general guidance the fatigue section of the bridge code [21] may be consulted. Readers wishing to learn something of the mechanics of fatigue are referred to the relevant section in McGuire [11].

1.5   CORROSION AND CORROSION PROTECTION

Steel readily corrodes (rusts) in moist air. Aggressive environments such as smoke, soot, sea water, acid or alkaline vapours will hasten the process. In a bad industrial area the rate at which the surface is 'lost' may reach 0.075 mm/year, more if particularly harmful agents such as sulphur dioxide, are present. Structural steelwork therefore needs to be properly protected [22]; guidance on this subject is provided in BS 5493.

The most common forms of protective treatment involve covering the exposed steel, either with paint or with a metallic coating, or possibly in the case of sheeting with a plastic coat. Concrete is not generally regarded as being capable of affording sufficient protection (except in the case of reinforcement).

Paint systems are described in CP 231. Generally a zinc or lead-based priming coat is applied first so as to provide a good foundation for the latter finishing coats. Care is necessary when using lead-based paints on account of their toxic nature; they should not be sprayed, applied in confined spaces or used on material that will subsequently be welded or flame cut.

Metallic coatings include galvanizing and sheradizing (both of which use zinc), electroplating, which is mainly confined to small items like fasteners, and metal spraying using either zinc or aluminium. Information on each of these techniques is provided in the relevant British Standard [23–26].

A common requirement for all schemes is cleanliness of the surface before treatment. For structural steelwork this is normally achieved by blast cleaning in which small abrasive particles such as iron are directed at the object using either compressed air or an impeller. Fabricating shops often arrange for incoming material to pass through the shotblasting plant on entry to the shop.

One alternative to the use of protective treatments consists of using a special corrosion-resistant steel which rapidly forms its own protective layer of oxide film. As shown in Fig. 1.13 this has the effect of reducing the corrosion rate to a negligible level after a few years. In Britain such materials are called 'weathering steels' [27], of which the best known is Cor-Ten. Originally developed by theUnited States Steel Corporation, this is now produced under licence in Britain by the BSC. Designs using weathering steel

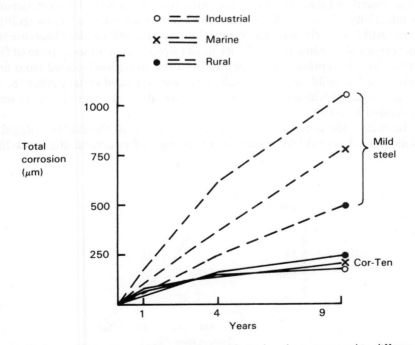

Fig. 1.13 Comparative corrosion rates for mild steel and cor-ten steel in different environments. (*After ref. 27.*)

clearly ought to exploit its particular properties; information on these is available [27].

It is important to appreciate that no coating is completely impermeable. Moreover, not surprisingly, there is a fair degree of correlation between the cost of a particular treatment and the degree of protection afforded by it. Therefore in common with most aspects of design the question of protection against corrosion is largely a matter of economics. To assist the designer, BS 5493 lists eight classes of environment (five external and three internal). Good designers will also try to 'design for prevention' by avoiding traps for dirt and moisture. A particularly useful presentation of the main aspects of corrosion protection for structural steelwork, covering corrosion prevention aspects of detailed design, surface treatment and protective systems is available in the ECSC guide [28]. A more limited discussion on minimizing the effects of corrosion is provided in reference [29], chapter 11.

1.6 FIRE PROTECTION OF STRUCTURAL STEELWORK

Although steel is an incombusible material, Fig. 1.14 shows how its strength may be reduced substantially by the action of high temperatures of the sort experienced in a major building fire. Moreover, because of its good thermal conductivity a bare steel beam may well assist in spreading a fire by igniting combustible material located beyond fire-resistant bulkheads. Therefore for most types of building the steelwork must be provided with some form of fire protection. Exceptions occur for single–storey structures isolated from any neighbouring buildings, some multistorey carparks and certain other 'zero-rated' buildings which can be shown not to be affected adversely by the heat generated by a fire.

In Britain the necessary requirements form part of the Building Regulations [30]. A useful interpretation for the case of structural steelwork has

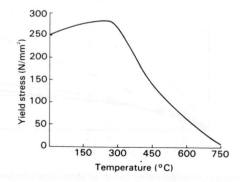

Fig. 1.14 Effect of elevated temperature on the strength of structural steel.

been prepared by Constrado [31]. The essential point is that sufficient protection must be provided for the main skeleton of the building to stand up long enough for people inside to escape. Thus minimum periods ranging from 30 minutes for a small residential building to 4 hours for a large store, are specified.

Such protection is afforded normally by encasing the steelwork in a suitable fire-resistant medium. In the so-called 'traditional method', brickwork, blockwork or concrete encasement is used. Since a typical thickness might be 50 mm for two hours' protection such methods tend to be slow and labour-intensive. An alternative would be wrapping in metal mesh which could subsequently be covered with a suitable plaster. Lightweight methods involve spraying the steelwork with some form of proprietory product. These have the advantage of lightness (thereby contributing little to dead load) and are usually less bulky. They also permit easier modification to the structure, an important consideration when it is remembered that one of the main attractions of a steel structure is the relative ease with which it can be altered at a later date. Included in these modifications could, of course, be changes in required fire resistance; increasing the thickness of the sprayed protection is a relatively easy undertaking. An assessment of 34 commercially available lightweight products, covering composition, appearance, method of fixing etc., is available from Constrado [32].

REFERENCES

1. British Steel Corporation. *Making iron*. Information Services Department, British Steel Corporation.
2. British Steel Corporation. *Open hearth furnace*. Information Services Department, British Steel Corporation.
3. British Steel Corporation. *Basic oxygen process*. Information Services Department, British Steel Corporation.
4. British Steel Corporation. *The electric arc furnace*. Information Services Department, British Steel Corporation.
5. British Steel Corporation. *Making steel*. Information Services Department, British Steel Corporation.
6. British Steel Corporation. *Guide to shaping processes in the steel industry*. British Steel Corporation.
7. British Steel Corporation. *Guide to structure and properties of steel*. British Steel Corporation.
8. Alpsten, G.A. *Variation in mechanical and cross-sectional properties of steel*. Swedish Institute of Steel Construction, Publication No. 42 (1973).
9. Nagaraja Rao, N.R., Lohrmann, M. and Tall, L. Effect of strain rate on the yield stress of structural steels. *ASTM Journal of Materials* 1 (1), 241–64 (1966).
10. Transport and Road Research Laboratory. *Recommended standard practices for structural testing of steel models*. TRRL Supplementary Report 254, Transport and Road Research Laboratory (1977).
11. McGuire, W. *Steel structures*. Prentice-Hall, Englewood Cliffs, New Jersey (1968).

12. Beedle, L.S. and Tall, L. Basic column strength. *ASCE J. of the Structural Division* **86** (ST7), 139–73 (July 1960).
13. Baker, M.J. *Variability in the strength of structural steels – a study in structural safety. Part 1: Material variability*. CIRIA, Technical Note 44 (September 1972).
14. Dwight, J.B., Chin, T.K. and Ratcliffe, A.T. Local buckling of thin-walled columns, Part 1: Effect of locked-in welding stress. CIRIA Research Report No. 12 (May 1968).
15. Young, B.W. Residual stresses in hot rolled members. *Proc. Int. Colloq. Column Strength*. IABSE, Zurich, 25–38 (1972).
16. Tall, L. *Structural Steel Design*. Ronald Press, New York, NY (1974).
17. Pratt, J.L. Introduction to the welding of structural steelwork. Constrado (April 1979).
18. Rolfe, S.T. and Barsoum, J.M. *Fracture and Fatigue Control in Structures*. Prentice-Hall, Englewood Cliffs, New Jersey (1977).
19. BCSA/Constrado. *Structural steelwork handbook*. London (1978).
20. Burdekin, F.M. *Materials aspect of BS 5400 Part 6: The design of steel bridges*, edited by Rockey, K.C. and Evans, H.R. Granada Publishing (1981).
21. British Standards Institution. *BS 5400 Part 3: Steel, concrete and composite bridges; Part 10: Code of practice for fatigue*. BSI, London (1980).
22. Constrado. *Protection of structural steelwork from atmospheric corrosion*. Constrado Publication 3/74 (December 1974).
23. British Standards Institution. *BS 729: Hot dip galvanised coatings on iron and steel articles*. BSI, London (1971).
24. British Standards Institution. *BS 4921: Sherardised coatings on iron and steel articles*. BSI, London (1973).
25. British Standards Institution. *BS 2569: Sprayed metal coatings*. BSI, London (1964 and 1965).
26. British Standards Institution. *BS 1706: Electroplated coatings of cadmium and zinc on iron and steel*. BSI, London (1960).
27. Chandler, K.C. and Kilcullen, M.B. Corrosion characteristics of weathering steels. Technical Note 10/CAB/TN/73, Corrosion Advice Bureau, British Steel Corporation (1973).
28. *Durability of steel structures*. European Coal and Steel Community (1982).
29. Davies, B.J. and Crawley, E.J. *Structural steelwork fabrication*. British Constructional Steelwork Association, Publication No. 7/1980.
30. *The building regulations 1976*. HMSO, London (1976).
31. Elliott, D.A. The building regulation 1976 – Application to steelwork of part E: Section 1: Structural fire precautions. Constrado (April 1980).
32. Elliott, D.A. *Protection of Structural Steelwork*, 2nd edn. Constrado (March 1981).

2
The basis of structural design

Structural design is an all-embracing term, which is used to cover general aspects of the subject, for example the choice of a particular structural form and a particular material, through the series of increasingly narrower decisions that leads eventually to points of detail such as the size of bolt required in a particular connection. Progress through each of these stages usually involves treating the problem in an increasingly quantitative manner. Although this book is concerned largely with the more detailed end of the process as it applies to steel structures, the material of this chapter should provide the reader with a taste of the wider aspects of the subject. Since BS 5950 is written principally, but not exclusively, with steel building structures in mind, the text concentrates on examples drawn from that area. Readers wishing to gain a wider appreciation of steel structures should therefore consult some of the references given in the Bibliography at the end of this chapter.

2.1 STRUCTURAL IDEALIZATION

Once the decision has been taken to construct a particular building in steel a suitable structural system must be selected. Factors which might influence the choice include:

(1) The spans involved – special consideration is necessary if there is a requirement for long spans or large, clear floor areas.
(2) The vertical loading – the presence of heavy point loads on floors or the need to accommodate cranes (*Cl. 2.4.1.2*).
(3) The horizontal loading – attention must be given to the way in which horizontal (wind) loading is to be resisted, for example by the framing

itself (by providing rigid joints), by bracing acting with the framing or by means of an independent bracing system such as a set of shear walls. This aspect of design is of particular importance for very tall buildings (*Cl. 2.4.2.3*).

(4) The services required – these include water, electricity and gas and are usually accommodated under the floors. In situations where large volumes of services are needed, as in hospitals, special forms of flooring permitting easy incorporation of the necessary pipework and ducting may be necessary.

(5) The ground conditions – clearly the type of ground on which the building is to be erected will dictate the form of foundations that must be used (pad, raft, piled, etc.) and this in turn must be taken into consideration when selecting the superstructure (*Cl. 2.4.2.4*).

Other items which might enter the discussion are the ways in which the building must be erected, accommodation of temperature effects (*Cl. 2.3*) and (if the steelwork is to be visible to the users such as the inside of the roof of an exhibition hall) the appearance. BS 5950 also requires steel-frame buildings to be tied together adequately and, in the case of multistorey buildings, to be capable of withstanding a limited amount of local damage without collapse (*Cl. 2.4.5*).

The way in which the designer decides to satisfy these requirements, several of which may well tend to conflict with one another, constitutes a difficult and frequently relatively neglected aspect of structural design. Its solution, which must draw heavily on experience of past satisfactory schemes, structural judgement, discussions with those other professions concerned with the design of the building as well as the client or user, knowledge of fabricating shop capabilities and erection techniques, etc., lies beyond the scope of this text. Wide reading of descriptions of actual projects (case studies), discussions with practising engineers, visits to fabricating shops and construction sites as well as a clear appreciation of structural behaviour all form part of the necessary educational process. In comparison with this aspect of design the actual proportioning of the members, detailed design of the connections, etc. is normally much more straightforward. However, a proper understanding of the more limited task is necessary before an engineer is competent to tackle the problem in its wider sense. Even when this stage has been reached greater experience and career advancement will cause the engineer to reconsider his definition of structural design as the boundaries of his involvement become wider.

The majority of steel buildings fit within one of the categories listed in Table 2.1. Of these, bearing wall construction, Fig. 2.1, in which the steel beams forming the roofs and floors bear directly on fairly substantial walls (usually constructed of brick or concrete blocks but sometimes of plain or reinforced concrete), is usually limited to low-rise, lightly loaded buildings such as schools.

A steel framework of beams and columns such as that shown in Fig. 2.2 is

Table 2.1
Broad categories of steel building construction

Type	Main use	Main considerations in design
Bearing wall	Low rise, lightly loaded	Structural design of steelwork is normally straightforward
Steel frame	Wide variety of types and size of building	'Simple construction' or 'continuous construction' depending on joint type used
Long span	Coverage of large column-free areas	Special forms of 'beam' may be required to span the required distances
High rise	Tall buildings, i.e. more than 20 storeys	Resistance to lateral forces due to wind load

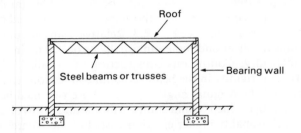

Fig. 2.1 Bearing-wall construction.

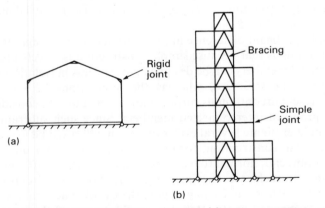

Fig. 2.2 Beam and column construction. (a) Portal frame — continuous construction. (b) Multistorey frame — simple construction.

much more common nowadays. Great versatility is possible, permitting this from of construction to be used for small, simple low-rise buildings as well as for much more complicated multistorey buildings. Depending on the type of beam-to-column joints employed, such systems are considered either as 'simple construction' (*Cl. 2.1.2.2*) or as 'continuous construction' (*Cl. 2.1.2.3*). For the former, rotation of the beams relative to the columns is

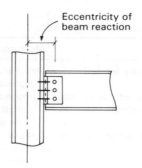

Fig. 2.3 Column moment due to eccentricity of beam reaction.

assumed to be possible so that beams may be designed as simply supported
with columns required to carry only those moments produced by the eccentricity of the beam reactions, see Fig. 2.3. Relatively simple connections may
be used to transmit shear and these can usually be bolted up in the field without undue difficulty. Continuous construction (also called 'rigid-frames')
assumes sufficient rigidity in the beam–column connections to maintain virtually unchanged the original angle between those two members when the
structure is loaded. Such connections will naturally involve additional
fabrication and probably higher erection costs but the greater rigidity produced in the structure due to its ability to develop flexural action may well
compensate in terms of reduced member sizes and the elimination of bracing.
This form of construction is very popular for low-rise industrial buildings of
the type shown in Fig. 2.2(a).

One very significant difference in the approach to the design of these two
types of framing is that because simple construction is effectively statically
determinate all members can be designed more or less in isolation in a single
pass through the structure, whereas the interactions between adjacent
members present in continuous construction necessitates the consideration of
at least a group of interconnected members. Since such subframes will be
statically indeterminate several cycles of design will often be necessary.

For long-span construction, for example roofs, the floor directly over a
hotel ballroom, etc., normal rolled sections may not have sufficient depth to
act as beams. In such cases they may be replaced by plate girders or trusses.
Coverage of very large areas may require the use of space frames, arches or
even cable-suspended roofs. Detailed consideration of these more exotic
forms of construction is beyond the scope of this text and the interested
reader is referred to the Bibliography.

For tall structures such as buildings of more than about 20 storeys depending on circumstances, microwave towers, etc., considerations of resistance to
lateral wind loading tend to dominate the design thinking. Figure 2.4 illustrates the two basic mechanisms for providing sway stiffness in a steel-frame
structure; either it can be braced, possibly using the internal walls, lift shafts,
etc., in which case adequate stiffness may be possible using main frames of

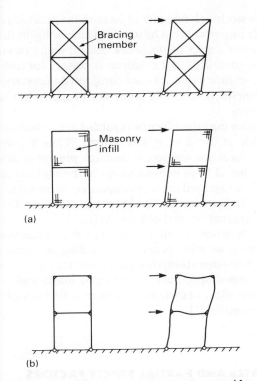

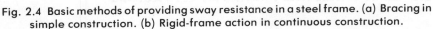

Fig. 2.4 Basic methods of providing sway resistance in a steel frame. (a) Bracing in simple construction. (b) Rigid-frame action in continuous construction.

'simple construction', or sway may be resisted by the inherent bending stiffness of rigid frame action. Various special systems have evolved to permit the construction of the 70–110 storey buildings that currently represent the world's tallest.

2.2 STRUCTURAL CODES

Much of the detailed information necessary for the design of steel structures is provided in codes of practice. In the context of this book the most important of these is BS 5950: The Structural Use of Steelwork in Building, Part 1: Code of Practice for Design in Simple and Continuous Construction. Chapters 3–7, each of which deals with the design of structural elements, will make frequent reference to the design procedures contained in that document. These cover items such as the relationship between strength and slenderness for a steel column, recommendations for the adequate spacing of holes for bolted joints and guidance on deflection limits. Whilst it is clearly necessary for the steelwork designer to be familiar with the provisions of this code, it is equally important that he uses it in an intelligent fashion. The code does not cover every aspect of steelwork design; many facets of the subject simply cannot be quantified in the manner necessary for codification, others are encountered so rarely that it is not considered necessary to lengthen the document by their inclusion, and some are properly left to textbooks on theory of structures.

A code of practice may therefore be regarded as a consensus of what is considered acceptable at the time it was written. Thus it contains a balance between accepted practice and recent research presented in such a way that the information should be of immediate use to the engineer in conducting his design. As such it is regarded more appropriately as an aid to design containing stress levels, design formulae and recommendations for good practice, rather than as a manual or textbook on design.

The steelwork designer will often need to refer to a number of other codes covering areas such as steel properties, welding of structural steelwork, properties of steel fasteners (bolts) and loads on structures as well as the other steelwork codes aimed specifically at bridges, masts and towers, offshore structures and steel silos. In certain cases he may find it useful to consult the codes of other countries [1].

2.3 LIMIT-STATES AND PARTIAL SAFETY FACTORS

Limit-states design simply provides the basic framework within which the performance of the structure can be assessed against various limiting conditions. When formulating procedures nowadays it is customary to do so in a way which recognizes the inherent variability of loads, materials, construction practices and approximations made in design; this usually involves the use of some concept of probability. The limiting conditions are normally grouped under two headings: ultimate or safety limit states and serviceability limit states. Table 2.2 lists those limit states which are usually considered relevant for structural steelwork. The attainment of one or more ultimate

Table 2.2
Limit states for structural steelwork

Ultimate (safety) limits – ULS	Serviceability limits – SLS
Overall loss of equilibrium (overturning)	Excessive deformation
Strength limits (general yielding, rupture, transformation into a mechanism, etc.)	Excessive vibration
Elastic or plastic instability	Corrosion
Fatigue (leading to fracture)	
Brittle fracture	

Table 2.3
Definition of basic limit-states terminology

Term	Definition
A limit state	A condition beyond which the structure would become less than completely fit for its intended use. If this happens, the structure is said to have entered a limit state.
The ultimate or safety limit state	Inability to sustain any increase in load.
The serviceability limit state	Loss of utility and/or requirement for remedial action.
Characteristic loads	Those loads which have an acceptably small probability of not being exceeded during the lifetime of the structure.
The characteristic strength of a material	The specific strength below which not more than a small percentage (typically 5%) of the results of tests may be expected to fall.
Partial safety factors	The factors applied to characteristic loads, and properties of materials to take account of the probability of the loads being exceeded and the assessed design strength not being reached.
The design load or factored load	The characteristic load multiplied by the relevant partial factor.
The design strength	The characteristic strength divided by the appropriate partial safety factor for the material.

limit states (ULS) may be regarded as an inability to sustain any increase in load. Serviceability (SLS) checks against the need for remedial action or some other loss of utility. Thus ULS are conditions to be avoided whilst SLS could be considered as merely undesirable. Since a limit-states approach to design involves the use of a number of specialist terms, simple definitions of the more important of these are provided in Table 2.3. A more detailed discussion of these and other matters relating to the general limit-states philosophy is provided in reference [2].

BS 5950 is not the first UK code to be based on this approach; it was preceded in 1972 by CP 110, the concrete code. Moreover, BS 5400 the bridge code, including Part 3 relating to the design of steel bridges, was prepared at much the same time as BS 5950, although it was actually published a few years previously. In other parts of the world limit-states steelwork codes are gradually appearing with the first of these having been published in Canada as long ago as 1974. In the UK work is in progress on the preparation of limit-states versions of the code for construction in each of the other main structural materials, namely timber, aluminium and masonry. Thus BS 5950 simply reflects the trend towards the general introduction of this more rational approach to structural design that is taking place for all the major construction materials on a worldwide basis.

Design for the ULS may conveniently be explained with reference to the type of diagram shown as Fig. 2.5. This compares the strengths R of a

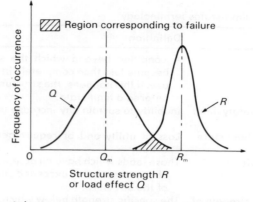

Fig. 2.5 Pictorial representation of the variability of loads and strengths.

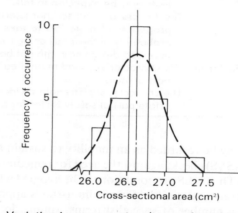

Fig. 2.6 Variation in cross-sectional area of steel beam sections.

number of nominally identical structures with the load spectrum Q that might be expected to occur during the lifetime of those structures. The fact that both quantities appear not as single vertical lines but as curves, termed frequency distributions, is in recognition of the variability not only of the loads experienced by a structure but also of the factors which influence the strength of the structure. Thus the load curve is broad, reflecting the variability of loading on a building structure, while the greater degree of control over its strength leads to a narrower strength curve. A simple illustration of the variability of structure strength R is provided by the data given in Figs. 2.6–2.8. These show how the naturally occurring variations in cross-sectional area and material strength of Figs. 2.6 and 2.7 (together with various other properties not illustrated) contributed to the spread of strengths shown in Fig. 2.8 when the beams were tested.

The shape of both the load and the strength curves of Fig. 2.5 will always be such that some small overlap will be present; this corresponds to a failure.

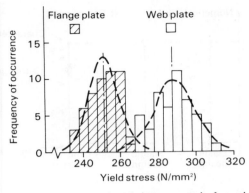
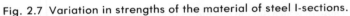

Fig. 2.7 Variation in strengths of the material of steel I-sections.

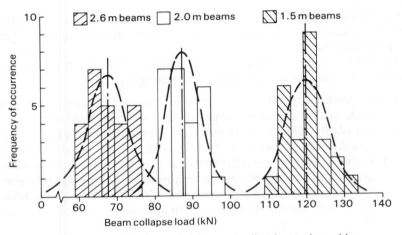

Fig. 2.8 Strengths of three sets of 25 nominally identical steel beams.

Good design consists of so proportioning the structure that this area corresponds to an acceptably small probability (say 1 in 100 000). In traditional allowable stress design this is achieved by scaling down the strength side of the design equation using a factor of safety γ_e as indicated in Fig. 2.9, while ultimate strength design compares actual structural strengths with the effects of factored-up loading by using a load factor of γ_p. Figure 2.9 shows how limit states design at the ULS employs separate factors on loading (γ_f) and strength (γ_m) in an attempt to cater for the different amounts of variability associated with these. Moreover, it is customary to break down the factors on each side into a number of partial safety factors, each of which reflects the degree of confidence in the particular contributing effect. Thus for a steel bridge for which the dead weight of the steelwork might be expected to be capable of more accurate assessment than the live loading due to traffic, the former will have a smaller partial factor associated with it than the latter. Typical figures might be 1.05 and 1.50 respectively. An internationally

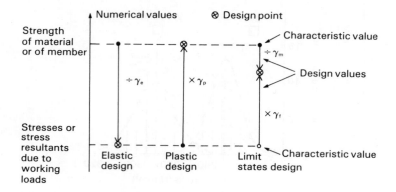

Fig. 2.9 Level at which design calculations are conducted for different approaches.

agreed list of γ-factors as they are called is available [2]. Actual numerical values are, however, usually decided upon in the code drafting committees of an individual country.

BS 5950 deliberately adopts a very simple interpretation of the partial safety factor concept in using only three separate γs:

(1) *Variability of loading* γ_l – Loads may be greater than expected; also loads used to counteract overturning may be less than intended.
(2) *Variability of material strength* γ_m – The strength of the material in the actual structure may vary from the strength used in calculations.
(3) *Variability of structural performance* γ_p – The structure may not be as strong as assumed in the design because of variations in the dimensions of members, variability of workmanship and differences between the simplified idealizations necessary for analysis and the actual behaviour of the real structure.

A value of 1.2 has been adopted for γ_p which, when multiplied by the values selected for γ_l, leads to the values of γ_f to be applied to the loading given in Table 2.4. Values of γ_m have been incorporated directly into the design strengths given. Thus the designer needs to use only the γ_f-values of Table 2.4 in his calculations. The different numerical values shown here are intended to provide approximately equal margins of safety under each form of loading.

Table 2.4
Values of γ_f to be applied to the loading – see Table 2 of BS 5950 for full list

Load type	Value of γ_f
Dead (maximum)	1.4
Dead (minimum)	1.0
Imposed (in the absence of wind)	1.6
Wind (acting with dead load only)	1.4
Wind and imposed (acting in combination)	1.2

2.4 LOADING

Assessment of the design loads for a structure consists of identifying the forces due to both natural and manmade effects which that structure must withstand and then assigning suitable values to them. Frequently several different forms of loading must be considered, acting either singly or in combination, although in some cases the most unfavourable situation might be easily identifiable. For buildings the usual forms of loading include dead load, live load, wind load, loads due to temperature effects and, in certain parts of the world, earthquake load. Other types of structure will each have their own special forms of loading, for example vehicle loading on highway bridges, fluid pressure inside storage tanks, and wave loading on marine structures.

When assessing the loads acting on a structure it is usually necessary to make reference to the appropriate codes of practice. Basic data on dead, live and wind loads for buildings in the UK are given in CP3, Chapter 5 [3], and BS 6399 [4] with more specialized information on matters such as the loads produced by cranes in industrial buildings (workshops, steel plants, etc.) being provided elsewhere [5]. For bridges and other special forms of structure the necessary loading data are normally provided in the code of practice appropriate to that type of structure [6, 7].

Determination of the dead load of a structure requires the estimation of the weight of the structure together with its associated 'non-structural' components. Thus, in addition to the bare steelwork (which strictly speaking should include items such as bolts and weld metal), the weights of floor slabs, partition walls, ceilings, plaster finishes and services (cable ducts, water pipes, etc.) must all be calculated. Since certain of these will not be known until after at least a tentative design is available, designers normally use approximations based on experience for their initial calculations. As an example, the weight of the steelwork in a light roof truss may be assumed as 50 kg/m². When the design is complete the actual dead load should be calculated; if it is significantly different from the assumed value then some modification of the design may be necessary. For the majority of steel buildings the weight of the actual steelwork will be less than 30% of the total dead load, so that quite large inaccuracies in its original assessment are unlikely to result in significant redesign.

The basis for estimation of live load is observation and measurement [8]. Live load in buildings covers items such as occupancy by people, office floor loadings, moveable equipment within the building, and machinery. Clearly different values will be appropriate for different forms of building – domestic, offices, warehouses, etc. The effects of snow, ice and hydrostatic pressure are normally included in this category.

Although the load produced on a structure by the action of the wind is really a dynamic effect, it is normal practice for most types of structure to treat this as an equivalent static load. Therefore, starting from the basic wind speed for the geographical location under consideration, suitably corrected to allow for the effects of factors such as topography, ground roughness and length of exposure to the wind, a dynamic pressure is determined. This is then converted into a force on the surface of the structure using pressure or force coefficients which depend on the building's shape. For some surfaces the final effect may well be to produce a negative suction force. The information contained in reference [3] is limited to the more usual building shapes; for larger and more complex arrangements the designer may require a model of his structure to be tested in a wind tunnel. Very tall buildings, high masts and suspension bridges often fall into this category.

The designer must also decide whether allowance is necessary for any temperature effects. These include expansion or contraction due to temperature difference, such as between the sunny and shaded parts of a bridge, as well as shrinkage and creep, as with concrete slabs.

REFERENCES

1. British Constructional Steelwork Association. *International Steel Handbook*. BCSA, London (1983).
2. Construction Industry Research and Information Association. *Rationalisation of Safety and Serviceability Factors in Structural Codes*. CIRIA Report 63 (July 1977).
3. British Standards Institution. *CP3: Chapter V, Loading, Part 2*. BSI, London (1972).
4. British Standards Institution. *BS 6399: Design Loading for Buildings, Parts 1, 2 and 3*. BSI, London (1984).
5. British Standards Institution, *BS 2573: Part 1: 1983, Rules for the Design of Cranes Part 1. Specification for Classification, Stress Calculations and Design Criteria for Structures*. BSI, London.
6. British Standards Institution. *BS 5400: Part 2: 1978. Steel, Concrete and Composite Bridges*. BSI, London.
7. British Standards Institution, *Lattice Towers: Loading*. Draft Code of Practice. BSI, London (1978).
8. Mitchell, G.R. and Woodgate, R.W. *Floor Loadings in Office Buildings – the Results of a Survey*. BRS Current Paper 3/71 (January 1971).

BIBLIOGRAPHY

McGuire, W. *Steel Structures*. Prentice-Hall, Englewood Cliffs, New Jersey (1968).
Gaylord, E.H. and Gaylord, C.N. *Design of Steel Structures* 2nd ed. McGraw-Hill, Kogakusha (1972).
Adams, P.F., Krentz, H.A. and Kulak, G.L. *Limit States Design in Structural Steel* 2nd ed. Canadian Institute of Steel Construction (1979).

British Steel Corporation. *Construction Guide*. BSC Sections, Redcar (1980).

Bresler, B. and Lin, T.Y. *Design of Steel Structures*. John Wiley & Sons, New York (1964).

Salmon, C.E. and Johnson, J.E. *Steel Structures* 2nd ed. Harpur & Row, New York (1980).

Makowski, Z.S. *Steel Space Structures*. Michael Joseph, London (1965).

Troitsky, M.S. *Cable Stayed Bridges, Theory and Design*. Crosby Lockwood Staples, London (1977).

Podolny, W. and Scalzi, J.B. *Construction and Design of Cable-Stayed Bridges*. John Wiley & Sons, New York (1976).

O'Conner, C. *Design of Bridge Superstructures*. Wiley-Interscience, New York (1971).

Heins, C.P. and Firmage, D.A. *Design of Modern Steel Highway Bridges*. Wiley-Interscience, New York (1979).

Gimsing, N.J. *Cable Supported Bridges, Concept and Design*. Wiley-Interscience, Chichester (1983).

Modern Steel Construction in Europe. Elsevier, Amsterdam (1963).

Constrado. *Building with Steel*. Approximately three issues per year each containing several articles on structural steelwork. Recent issues have featured:

Building with Steel 7 (1978–9)
1. Steel reaches out
2. Steel in architecture
3. Steel and fire
4. Steel and commerce
5. Industrial buildings
6. Cladding

Building with Steel 8 (1980–81)
1. The art of the fabricator
2. Protection of steelwork
3. Multi-storey buildings
4. Bridgework

Building with Steel 9 (1982–4)
1. Steel in housing
2. Fire protection
3. Corrosion protection
4. Multi-storey buildings
5. Steel and transport
6. Sport and leisure

Acier-Stahl-Steel. Four issues per year each containing several reports of steelwork projects.

3
Tension members

Tension members are used quite frequently in a variety of steel structures; some of these uses are illustrated in Fig. 3.1. Depending principally upon the magnitude of the load to be carried and the type of interconnection to be used between members, any of the structural sections shown in Fig. 3.2 may be suitable. Although the major design consideration will be the provision of adequate tensile strength, some limitation on slenderness is usually also necessary in order to eliminate possible problems due to excessive sag under

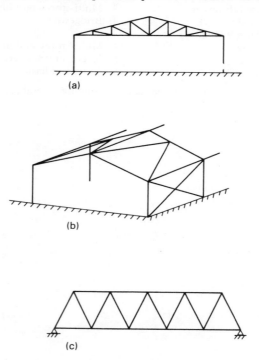

(a)

(b)

(c)

Fig. 3.1 Structures containing tension members. (a) Roof truss; (b) bracing for a portal frame building; (c) bridge truss.

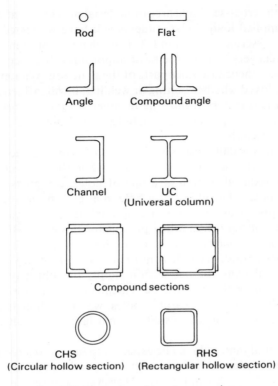

Fig. 3.2 Examples of tension members.

self-weight [1], flutter due to wind loads or vibration caused by moving loads. For this reason rods or flats are of limited use, especially if required to act in compression due to reversal of load. When used as diagonal bracing, rods may be pretensioned so as to reduce their self-weight deflections [1].

Angles, used either singly or in pairs placed back to back, are suitable for many applications; two of the more visible examples are in small to medium roof trusses or in transmission towers. When heavy loads have to be carried over long spans, as in a truss bridge, then larger rolled sections, possibly acting in combination, may be necessary. Tubes, either circular or rectangular, may be used as bracing or as the main members in trusses or space frames; care is necessary in deciding upon the jointing arrangements [2].

3.1 BEHAVIOUR OF MEMBERS IN TENSION

The design of a member subjected to a tensile force is probably the most straightforward of all structural design problems. Essentially it consists of ensuring that the cross-sectional area of material provided is at least adequate to resist the applied load. Most students will have witnessed the standard laboratory test described in section 1.3. The behaviour of a tension member is in many respects very similar, the most important difference being that the member will be attached to other parts of the structure. Whatever method of jointing is employed whether bolting or welding, it will influence the manner in which load is transferred into the member. In the case of fastening by mechanical means the presence of the holes will also have a direct effect on the member's strength.

This problem is usually discussed in terms of gross and net sections. The former is simply the original cross-section while the net section is usually defined as the reduced section at a line of holes, i.e. gross section minus allowance for holes. The effect of a hole in a tension member amounts to more than simply the absence of some material. In the immediate vicinity of the hole a stress concentration will be present and this will itself be affected by the localized force applied by the fastener. However, because of the ductility possessed by structural steels, it is normal in design to neglect these other effects and to calculate the net section simply by subtracting the area of the hole(s). In doing this it must be remembered that most types of bolt (see section 7.2) are used in clearance holes, where the hole is made slightly larger than the bolt diameter, usually 2 mm larger for bolt diameters up to 24 mm.

Since removal of material may be expected to have a weakening effect one might expect that failure would normally occur at the smallest net section i.e. across the line of holes AA in Fig. 3.3. However, because it is desirable that failure occur in a ductile rather than a brittle manner, it is usual to try to ensure that the gross section yields before the ultimate tensile strength of the net section is reached [3]. This greatly increases the amount of deformation that the member can sustain and consequently gives a better indication of impending failure.

Even for a connection between two flat bars of the type shown as Fig. 3.3 some eccentricity of the line of action of the tension T will be present. However, providing this is small the resulting bending effects will be such that their influence on the member's ultimate strength may be neglected. Thus BS 5950 allows certain types of tension member to be designed for tension only,

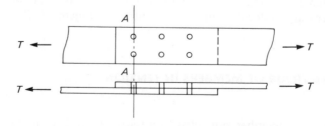

Fig. 3.3 Tearing at a line of holes.

makes safe approximate allowances in other cases, and only occasionally requires the explicit consideration of bending effects.

3.2 BASIC DESIGN APPROACH

3.2.1 Effective section

By ensuring that the ratio of net area A_n to gross section A exceeds the ratio of yield strength Y_s to ultimate tensile strength U_s, BS 5950 effectively allows design to be based on the condition of yield of the gross section. Thus the effective area at a connection A_e is defined in *Cl. 3.3.3* as K_e times the net section, where K_e adopts the values 1.2, 1.1 and 1.0 for steel grades 43, 50 and 55 respectively, with the limitation that the effective area cannot exceed the gross section. The tension capacity P_t of the member is therefore given by

$$P_t = A_e p_y \qquad (3.1)$$

in which p_y is the design strength of the steel obtained from *Table 6*. Reference to this table shows that it differentiates between the three basic grades of structural steel Gr. 43, 50 and 55; it also recognizes the smaller differences in p_y that result from the different manufacturing processes used for the different types of structural member and it allows for the gradual reduction in material strength that results from the use of thicker material.

3.2.2 Net section

Where holes are arranged in parallel rows at right angles to the member axis as shown in Fig. 3.3 the net section is obtained by subtracting the maximum sum of the hole areas across any cross section from the gross area, i.e.

$$A_n = A_g - \Sigma td \qquad (3.2)$$

in which t = plate thickness and d = hole diameter

Example 3.1
A flat bar 200 mm wide × 25 mm thick is to be used as a tie. Erection considerations require that the bar be constructed from two lengths connected together with a lap splice using six M20 bolts as shown in Fig. 3.3. Calculate the tensile strength of the bar assuming steel of design strength 265 N/mm².

Solution
Hole clearance = 2 mm
Gross section   = 200 × 25 = 5000 mm²
From eqn (3.2), Net area AA = 200 × 25 − 2 × 22 × 25 = 3900 mm².
From *Cl. 3.3.3* for grade 43 steel effective area = 1.2 × 3900 = 4680 mm², which is less than the gross area.
From eqn (3.1), P_t = 265 × 4680 N = 1240 kN

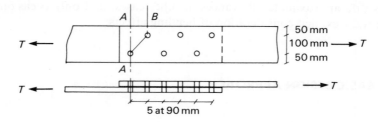

Fig. 3.4 Zig-zag failure mode for staggered holes.

Inspection of Example 3.1 reveals that the effective section is some 94% of the gross section. Over most of the member's length the section is therefore overdesigned, i.e. its capacity exceeds the required strength. This will usually be the case when parallel rows of holes are present. However, it is possible to reduce or even to eliminate this overdesign by staggering the holes as shown in Fig. 3.4. This introduces the possibility of failure occurring at either of the two net sections AA across the plate or AB in a zig-zag. Both sections should normally be checked.

Calculations of the net section at a line of staggered holes is covered in *Cl. 3.4.3*. This makes some allowance for the slightly increased strength corresponding to the zig-zag mode, by reducing the amount of material considered as ineffective to the total hole area along the section less a factor, to give

$$A_e = A_n + \frac{S_p^2 t}{4g} \tag{3.3}$$

in which t = thickness of plate and S_p and g are the staggered pitch and gauge as shown in Fig. 3.5.

Example 3.2
Repeat Example 3.1 for the new arrangement of holes shown in Fig. 3.4.

Solution
From eqn (3.2), net section AA = $200 \times 25 - 22 \times 25 = 4450$ mm²
From eqn (3.3), net section

$$AB = 200 \times 25 - 2 \times 22 \times 25 + \frac{90^2 \times 25}{4 \times 100} = 4406 \text{ mm}^2$$

Minimum net section is AB and the effective area is therefore $1.2 \times 4406 = 5287$ mm² which exceeds the gross section.
Take effective area as 5000 mm² and from *Cl. 3.1.1*
Tensile strength = 265×5000 N = 1325 kN

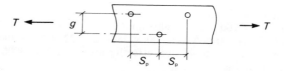

Fig. 3.5 Definition of gauge g and staggered pitch S_p.

Thus staggering the holes results in a situation where design is governed by the condition of yield of the gross section with no loss of efficiency.

3.3 ECCENTRIC CONNECTION

Although it is usually regarded as 'good practice' to try to ensure that load is transmitted into a tension member so that it acts along the member's centroidal axis, this will not always be possible. One obvious example would be a single angle for which the centroidal axis lies outside the cross-section and connection to one or other leg would clearly introduce an eccentricity. For certain types of member, however, the moments produced by these eccentricities are relatively small and it is not actually necessary either to calculate them or to make explicit allowance for them in design. Rather, the effective area may be reduced slightly as shown in Fig. 3.6 so that that part of the member's capacity which is not now being used to carry axial load is available to withstand the bending [4]. The justification for this approach is quite simply that, providing the correct sort of reduction in effective area is made, then it can be shown to provide good estimates of the strengths of single angles with either welded [4] or bolted [5] gusset plate connections on one leg.

Due to the eccentricity of load application the initial tendancy as such a member is tested is for the gussets to deform so as to enable the line of action of the applied load to approach the centroidal axis of the angle, as illustrated in Fig. 3.7. Thus at a load of about 50% of ultimate [4], while strains near the centre will be approximately uniform over the cross-section, sufficient bending will have occurred near the ends for yield of the attached leg to have started. This load may be determined approximately as that which just causes yield assuming the tension to act at the midplane of the gusset. Taking c as the

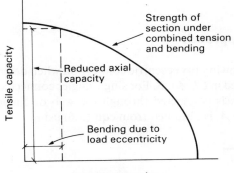

Fig. 3.6 Use of reduced axial capacity to allow for interaction of tension and bending.

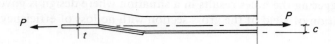

Fig. 3.7 Bending of gusset to permit reorientation of load in angle connected to one leg.

distance between the centroid of the angle and the extreme fibre in contact with a gusset of thickness t, this gives a moment of $P(c + t/2)$. Rearranging for the value of P at which first yield occurs gives

$$P_y = A\sigma_y \left(\frac{1}{1 + AC(c + t/2)/I} \right) \qquad (3.4)$$

in which I = second moment of area of the angle about an axis parallel to the gusset.

Further loading will eventually produce full section yield in the central region, leading to large deformations until eventual failure by fracture of the angle in the connected region. The authors of both references [4] and [5] recommended that design be based on the load corresponding to the commencement of large deformations and suggested an expression of the form

$$P = \sigma_y (a_1 + fa_2) \qquad (3.5)$$

in which a_1 = net area of connected leg

a_2 = gross area of outstanding leg
f = factor to allow for bending effects

Thus the effective area A_e of the section is

$$A_e = a_1 + fa_2 \qquad (3.6)$$

Use of eqn (3.5) was shown to provide quite consistent predictions of capacity.

3.3.1 Design approach

The method of allowing for eccentric connection when determining effective areas A_e is explained in *Cl. 4.6.3*. For single angles connected through one leg only, single channels connected through the web or single tees connected through the flange A_e is obtained from eqn (3.6) using

$$f = \frac{3a_1}{3a_1 + a_2} \qquad (3.7)$$

When two identical parallel components are in contact back-to-back or are separated by a small gap with regular and frequent interconnection, eqn (3.6) may still be used providing f is taken as

$$f = \frac{5a_1}{5a_1 + a_2} \tag{3.8}$$

Example 3.3
Determine the axial strength of a $150 \times 90 \times 15$ mm angle section when it is used as a tie, the end connection being a single 16 mm bolt through the longer leg. Assume steel of design strength $p_y = 450$ N/mm².

Solution
From BCSA tables, $A = 150$ mm, $B = 90$ mm, $t = 15$ mm
Gross area of non-attached leg $a_2 = 15 \left(90 - \frac{1}{2} \times 15\right) = 1237$ mm²
From eqn. 3.2, net area of attached leg $a_1 = 15 \left(150 - \frac{1}{2} \times 15\right) - (16 + 2) \times 15$
$$= 1867 \text{ mm}^2$$
From eqns (3.6) and (3.7), $A_e = a_1 + fa_2$
$$= 1867 + \left(\frac{3 \times 1867}{3 \times 1867 + 1237}\right) 1237 = 2800 \text{ mm}^2$$
From eqn (3.1), $P_t = 450 \times 2880$ N $= \underline{1296 \text{ kN}}$

This compares with a member strength (no allowance for holes or eccentricity) of 1518 kN, i.e. a loss of 15%. This suggests that when selecting an initial trial section to check whether it will be adequate to resist the design load, a member whose area exceeds that given by (design load)/(design strength) by about 15–20% should be chosen. The 'extra capacity' should then be enough to balance the necessary allowances for holes and eccentricity. If welded end connections are to be used the margin should be reduced to about 10% since only eccentricity is involved. Clearly the exact amount that will be 'lost' depends on the relative areas of the connected and unconnected parts of the section.

Example 3.4
Select a suitable equal angle section to carry a tensile force of 900 kN assuming (i) single M16 bolted end connections, (ii) welded end connections. Assume steel of design strength $p_y = 275$ N/mm².

Solution
(i) Approximate required area $= 1.2 \times 900 \times 10^3/275 = 4000$ mm²
From BCSA handbook, nearest section is $150 \times 150 \times 15$ mm which has an area of 4300 mm²
Gross area of non-attached leg
$$a_2 = 15 \left(150 - \frac{1}{2} \times 15\right) = 2137 \text{ mm}^2$$
From eqn (3.2), net area of attached leg
$$a_1 = 15 \left(150 - \frac{1}{2} \times 15\right) - (16 + 2) \times 15 = 1867 \text{ mm}^2$$
From eqns (3.6) and (3.7), $A_e = a_1 + fa_2$
$$= 1867 + \left(\frac{3 \times 1867}{3 \times 1867 + 1237}\right) 2137 = 3414 \text{ mm}^2$$
From eqn (3.1), $P_t = 275 \times 3414$ N $= \underline{938 \text{ kN}}$, which is satisfactory.

(ii) Approximate required area $= 1.1 \times 900 \times 10^3/275 = 3600$ mm²
From BCSA handbook, nearest section is again $150 \times 150 \times 15$ mm which has an area of 4300 mm²

Gross area of non-attached leg $a_2 = 15(150 - \frac{1}{2} \times 15) = 2137$ mm^2
Since no deduction is necessary for connection by welding this is also the value of a_1.

From eqns (3.6) and (3.7), $A_e = 2137 + \left(\dfrac{3 \times 2137}{3 \times 2127 + 2137}\right) 2137 = 3740$ mm^2

From eqn (3.1), $P_t = 275 \times 3740$ N $= \underline{1029\ \text{kN}}$, which is satisfactory.

Checking the next section down (150 × 150 × 12 mm) its strength (welded connection) is only 832 kN so that in this case the type of end connection employed does not affect the choice of section. However, if an unequal leg angle is acceptable then a 200 × 150 × 12 mm section having a gross area of 4080 mm^2 would carry the load. Because a_1 and a_2 in eqn (3.6) for such a section would not now be equal the value of P_t would depend on which of the legs is connected. If it is the longer leg, $P_t = 1021$ kN, a figure that reduces to 617 kN if the larger leg is only partly ($f = 0.69$) effective.

REFERENCES

1. Kitipornchai, S. and Woolcock, S.T. *Design of Diagonal Roof Bracing Rods and Tubes*. University of Queensland, Department of Civil Engineering, Report No. CE52 (June 1984).
2. CIDECT. *The Strength and Behaviour of Statically Loaded Welded Connections in Structural Hollow Sections*. Monograph No. 6, Comité International pour le Développement et L'Étude de la Construction Tubulaire (1981).
3. Adams, P.F., Krentz, H.A. and Kulak, G.L. *Limit States Design in Structural Steel – SI Units*. Canadian Institute of Steel Construction (1979).
4. Regan, P.E. and Salter, P.R. Tests on welded-angle tension members. *Structural Engineer* **62B** (2), 25–30 (1984).
5. Nelson, H.M. *Angles in Tension*. Publication No. 7, British Constructional Steelwork Association, 9–18 (1953).

EXAMPLES FOR PRACTICE

1. Select the lightest square hollow section from the *Structural Steel Handbook* in Grade 50 steel capable of carrying a factored axial tensile load of 730 kN, assuming that full-strength welded end connections are provided.

[90 × 90 × 6.3 mm]

2. Determine the tensile capacity of an 80 × 80 × 8 mm angle section in Grade 43 steel assuming that it contains a splice in which cover plates are provided to both legs. Assume the use of one row of M20 bolts in each leg arranged in pairs, i.e. not staggered.

[290 kN]
[241 kN]

3. Determine the tensile capacity of the flat bar tie in the arrangement shown in Fig. 3.8 assuming Grade 43 steel and M20 bolts.

[1272 kN]

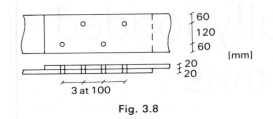

Fig. 3.8

4. Determine the tensile capacity of the flat bar tie in the arrangement shown in Fig. 3.9 assuming Grade 50 steel and M20 bolts.

[2394 kN]

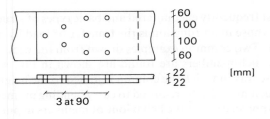

Fig. 3.9

5. Show that a 100 × 100 × 12 mm equal angle section in Grade 43 steel is capable of carrying an axial tension of 450 kN assuming the use of a welded connection on one leg only.

[543 kN]

6. Determine the tensile capacity of a 150 × 90 × 10 mm angle in Grade 43 steel assuming:

(a) Connection through the longer leg by 2 rows of M20 bolts

[460 kN]

(b) Connection through the shorter leg by one row of M24 bolts

[381 kN]

7. Determine the tensile capacity of a pair of 150 × 75 × 12 mm angles having the long legs back to back, assuming that 2 rows of M20 bolts are used and that the steel is Grade 43.

[1214 kN]

4

Axially loaded columns

One of the most frequently encountered and basic types of structural member is the column whose main function is the transfer of load by means of compressive action. Two common examples drawn from the wide range of structures in which such members are found are shown in Fig. 4.1. Depending upon the precise way in which the column is joined to the neighbouring parts of the structure, it may also be required to carry bending moments. Nevertheless, a proper appreciation of the behaviour of members in pure compression forms an important first step in understanding this more general problem because design for combined loading (considered in Chapter 6), i.e. compression and bending, is usually based upon considerations of the interaction of the various individual load components.

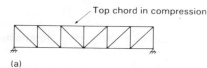

(a)

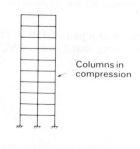

(b)

Fig. 4.1 Examples of compression members. (a) Compression members in a truss; (b) compression members in a building frame.

The response of a compression member to a nominally axially applied load depends upon a number of factors, the most important of which are its length and cross-sectional shape, the characteristics of the material from which it is made, the conditions of support provided at its ends and the method used for its manufacture. Table 4.1 lists the major forms of response.

4.1 STOCKY COLUMNS

4.1.1 Stub column behaviour

The results of a typical laboratory compression test on a short length of rolled section are shown in Fig. 4.2 in the form of a load versus end-shortening curve. Such a test is often referred to as a 'stub column test'. Comparison with the results of a compression test on a small coupon cut from the cross-section (presented previously as Fig. 1.6) shows that the major difference between the two is the lower limit of proportionality exhibited by the test on the full cross-section. The explanation for this lies in the non-uniform yielding of the stub column caused by the presence of residual stresses [1]. Thus those fibres which contain residual compression have their effective yield point reduced, while those containing residual tension have theirs increased. In both cases, however, the full strength of the material can be achieved with the stub column failing at its squash load. Although the actual collapse of the stub column would normally be precipitated by local buckling, for compact sections this would not occur until after considerable plastic straining had taken place. Many thousands of stub column tests have now been conducted and these demonstrate quite conclusively that the appropriate basis for the design of stocky columns of compact cross-section is the squash load.

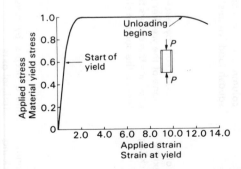

Fig. 4.2 Stress–strain behaviours of a full section in compression — stub-column response.

Table 4.1
Possible failure modes for an axially loaded column

Mode	Description	Illustration	Section	Comments
Squashing	Providing the length is relatively small (stocky column) and its plate elements are not too thin (compact cross-section) then the column will be capable of attaining its squash load (yield stress × area)		4.1	Member needs to be extremely stocky
Overall flexural buckling	Failure occurs by excessive deflection in the plane of the weaker principal axis, the load at which this occurs becoming progressively less as the column slenderness is increased		4.2	Controls the design of most compression members
Torsional buckling	Failure occurs by twisting about the longitudinal axis			This mode is unlikely for hot-rolled sections or for fabricated sections of 'normal' proportions but may be important for lighter cold-formed members, particularly unsymmetrical shapes.

Local
buckling

Failure occurs by buckling of one or more individual plate elements, e.g. flange or web, with no overall deflection; this may be prevented by placing suitable limits on plate width-to-thickness ratios; alternatively, where such limits are exceeded, the design strength must be reduced

4.1.2   Proportions of normal hot-rolled sections are such as to preclude this in most instances; needs to be considered for fabricated members or cold-formed sections.

Local
failure

Where compound members are formed by joining together two or more shapes to form a lattice cross-section, failure of a component member may occur if the joints between members are too widely spread.

4.4   Design rules usually require intermediate fastening to be sufficient to permit design for overall buckling as an equivalent solid strut.

4.1.2   Local buckling in columns

Not all stocky columns will be capable of attaining their full squash load. If the individual plate elements which make up the cross-section, for example the web and the two flanges in the case of an I-section, are thin then local buckling of the type shown in Fig. 4.3 may occur at a lower load. Analysis of this type of failure is somewhat complex so design rules are based largely on experimental data. For columns (which is the only case dealt which here – see Chapter 5 for more details) it is frequently possible to simply 'design out' the problem by so limiting the proportions of the component plates that local buckling effects will not influence the cross-section's strength. In cases where more slender plating is to be used the section's strength must be suitably reduced.

4.1.2.1   Design approach

Cl. 3.5.2 of BS 5950 classifies those sections for which yield may be attained without prior local buckling as semi-compact. Upper limits for this range are:

Flanges, i.e. plate elements supported along one longitudinal edge	Webs, i.e. plate elements supported along both longitudinal edges
$b/T \ngtr 15\sqrt{(275/p_y)}$	$b/T \ngtr 39\sqrt{(275/p_y)}$ for a rolled section
$b/T \ngtr 13\sqrt{(275/p_y)}$	$b/T \ngtr 28\sqrt{(275/p_y)}$ for a welded section

where the method to be used to assess *b*, *d* and *T* is given in *Figure 3*. Stricter limits are imposed for welded plates in recognition of the weakening effect of the more severe residual stress present [2]. Sections which do not meet these limits are classified as 'slender' and assessment of their load-carrying

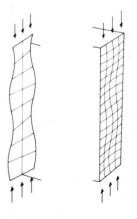

Fig. 4.3 Local buckling in box and I-section columns (deformations of a single flange only shown). (*BSC Teaching Project, Imperial College, 1985.*)

capacity must reflect the influence of local buckling. *Cl. 3.6.3* suggests one way of treating this in which a reduced design strength p_y for which the section would just be semi-compact is used throughout the calculations of member strength. However, relatively few rolled sections are affected when using other than the higher grades of steel. In particular, no UC sections in Grade 43 steel are less than semi-compact.

Example 4.1

A 305 × 102 mm UB33 is to be used as a short column carrying axial load. Is its compressive strength likely to be affected by local buckling assuming (i) Grade 43 steel, (ii) Grade 50 steel?

Solution
From section tables, $B = 102.4$ mm, $T = 10.8$ mm, $d = (275.8 - 2 \times 7.6)$ mm, $t = 6.6$ mm, $A = 40.8$ cm³.
Reference to *Figure 3* shows that T, d and t correspond to those used in *Table 7*.

$b = \frac{1}{2}(102.4 - 6.6 - 2 \times 6.6) = 41.3$ mm
$b/T = 41.3/10.8 = 3.82$

From *Table 7*, for $p_y = 275$ N/mm² limit is 15

$d/t = 260.6/6.6 = 39.5$

From *Table 7*, for $p_y = 275$ N/mm² limit is 39 (negligible excess)
 Full cross-section is available and $P_c = 275 \times 4080$ N
$$= \underline{1122 \text{ kN}}$$
From *Table 7* for $p_y = 355$ N/mm², flange limit is $15\sqrt{(275/355)} = 13.2$
From *Table 7* for $p_y = 355$ N/mm², web limit is $39\sqrt{(240/340)} = 34.3$
Therefore web d/t limit is exceeded. From *Cl. 3.6.3* use a reduced design strength of

$$355 \times [39.51\sqrt{(275/355)} - 8] = 298 \text{ N/mm}^2$$

$P_c = 298 \times 4080 = \underline{1216 \text{ kN}}$

In this case local buckling reduces the compression strength by about 15%. Since the web proportions control and most of the section's area is concentrated in the (semi-compact) flanges, a more rigorous allowance for the reduced effectiveness of the web only should lead to a much smaller loss of design capacity.

Example 4.2

Check whether the welded column section shown in Fig. 4.4 could be designed for its full squash load. Assume Gr. 50 steel.

Solution
From *Table 7* the flange limit is $13\sqrt{(275/340)} = 11.7$
Actual b/t, noting how b is defined in *Figure 3*, $= (200 - 12.5)/20 = 9.38$
Web limit from *Table 7* is $28\sqrt{(275/340)} = 25.2$
Actual d/t, noting how d is defined in *Figure 3*, $= 400/10 = 40$
Web is slender. Therefore either reduce design strength accordingly or replace 10 mm web by one of at least $(400/25.2) = 15.9$ mm; use 16 mm web and design for full squash load.

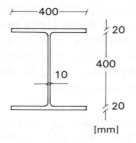

Fig. 4.4 Column cross-section of Example 4.2.

4.2 SLENDER COLUMNS

4.2.1 Background to the problem

Although the theory of the elastic stability of perfect pin-ended struts [3, 4], sometimes referred to as the Euler theory, provides some insight into the behaviour of slender compression members, it omits the consideration of a number of important factors [5]. These are often grouped under the general heading of 'imperfections' and include such factors as initial lack of straightness, accidental eccentricities of loading, residual stresses and variation of material properties over the cross-section. Their combined effect is to produce the type of relationship between theory and experiment shown in Fig. 4.5. Thus, while very slender columns fail at loads which are close to their elastic critical load, columns of intermediate slenderness (which account for a large proportion of cases found in actual construction) collapse at loads some way below either the elastic critical load or the squash load. Only by resorting to complex numerical methods is it possible for an analysis to include the effects of these imperfections.

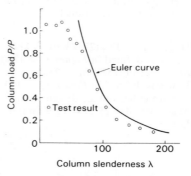

Fig. 4.5 Typical column test data compared with basic Euler strut theory, data on high-strength H-sections, from ref. 6.

The design, as opposed to the analysis, of columns is usually based on the concept of one or more 'column curves' which give load-carrying capacity directly as a function of slenderness. Figure 4.6 presents the set of four curves used in BS 5950. These have been based on the careful study [7–10] of both theoretical and experimental data. The reason for using more than one curve becomes clear when data of the form shown in Fig. 4.7 are examined. Despite the inevitable scatter associated with column tests the results indicate clear differences in strength between columns of the same slenderness but different type. This is largely a consequence of the different ways in which progressive yielding affects the stiffness of the various shapes; a factor that is itself dependent upon the pattern of residual stresses present. This in turn depends upon the method of manufacture which will also influence other controlling factors such as straightness and dimensional tolerances. Thus, in common with other recent national codes, BS 5950 recognizes this fact by requiring the use of different column curves for different classes of column.

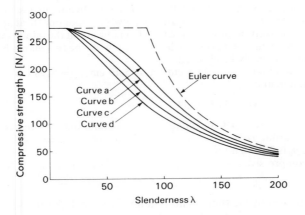

Fig. 4.6 Column design curves of BS 5950, $P_y = 240$ N/mm^2.

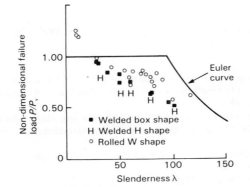

Fig. 4.7 Experimental data for the column strength of different types of steel section. (*Chen and Atsuta, Theory of Beam Columns, vol. 1. McGraw-Hill 1976, by permission.*)

4.2.2 Design approach

A formula describing the four curves of Fig. 4.6 is presented in *Appendix C* of BS 5950. However, it is not necessary to use this in actual design (unless column design is to be programmed) since tables of design axial strength p_c versus slenderness $\lambda = \ell/r_{min}$, in which r_{min} is the minimum radius of gyration, for each curves are given for a complete range of yield strengths p_y in *Table 27*. The particular table, i.e. whichever column curve should actually be used, must first be ascertained by reference to the selection table, *Table 25*. The following worked examples illustrate the process.

Example 4.3
Calculate the compressive resistance of a 203 × 203 mm UC60 of height 3.1 m. Assume that the conditions at both ends in the *xx* and *yy* planes are such as to provide 'simple support'. Take the design strength of the steel p_y as 275 N/mm².

Solution
Unless the axis about which buckling will occur is obvious all possibilities must be checked. For UC sections r_y is normally between about one third and one half of r_x so that the likely mode of failure is by buckling about the minor axis. However, in cases where different effective lengths apply for the two planes both possibilities should normally be checked.
From section tables, $A = 75.8$ cm², $r_y = 5.19$ cm, $r_x = 8.98$ cm.
Work in mm and N.

$$\lambda = l/r_y = 3100/51.9 = 59.7$$

From *Table 25* for a UC buckling about the minor axis, curve c is appropriate. Therefore from *Table 27c* for $\lambda = 59.7$ the corresponding value of the axial strength p_c is 201 N/mm².
Hence compressive resistance $P_c = 201 × 7580 = 1524 × 10^2$ N
$$= \underline{1524 \text{ kN}}$$

Clearly for this example there is no real need to check for buckling about the major axis since $r_x > r_y$. It is left to the reader to show that this is indeed the case by using column curve b to find that $P_c = 1948$ kN.

Example 4.4
Repeat the previous example for the 200 × 200 × 59.9 mm equal leg angle section shown in Fig. 4.8.

Solution
Since the principal axes for an angle section do not coincide with the rectangular *x–x* and *y–y* axes the buckling strength about the minor principal axis *v–v* should normally be checked.
From *Table 25*, curve c is appropriate for buckling about any axis.
Therefore only the axis about which the slenderness is greatest need be considered.
From section tables, $r_{xx} = r_{yy} = 6.11$ cm, $r_{vv} = 7.79$ cm
Work in mm and N.
Max. $\lambda = 3100/39.2 = 79.1$
From *Table 27c* for $\lambda = 79.1$ and $p_y = 275$ N/mm².
$P_c = 163$ N/mm².
Hence $p_c = 163 × 7630 = 1244 × 10^3$ N $= \underline{1244 \text{ kN}}$

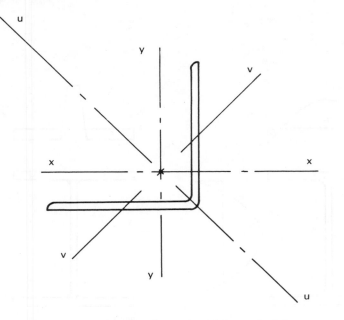

Fig. 4.8 Equal leg angle section of Example 4.6.

This is approximately 17% less than the strength of the UC section of almost identical weight. This is a direct result of the less favourable arrangement of material with regard to bending stiffness leading to a lower value for r_{min}. However, as explained in section 4.3 angles are frequently used as compression members in lightly loaded trusses because of the relative ease of making connections between them.

4.2.3 Welded sections

The column curves of Fig. 4.6 are intended for application to hot-rolled shapes. Available data for welded shapes [5, 10, 12] show that because of the rather different pattern of residual stresses present (see Fig. 4.9), a column curve of a slightly different shape should be used. Rather than increase the number of column curves still further, *Cl. 4.7.5* of BS 5950 deals with this problem by the simple expedient of requiring welded columns to be designed as if their yield strength was $p_y = 20$ N/mm². This device leads to the correct sort of design strengths over much of the range [7] although it does, of course, produce an inconsistency for very stocky columns which cannot be designed for their full squash load.

In cases where I- or H-sections are welded together from flame-cut plates the effect of the flame-cutting will be to produce beneficial tensile residual stresses at the flange tips as shown in Fig. 4.9(c). *Tables 25 and 26* therefore permit design to be based on the full value of p_y in such cases.

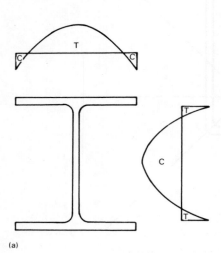

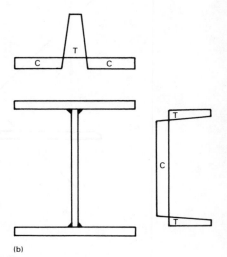

(a)   (b)

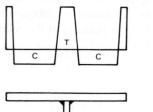

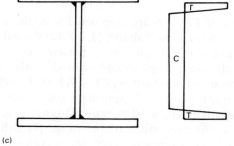

(c)

Fig. 4.9 Typical residual stress patterns in column sections made by different processes.

Example 4.5

A heavy column is required to support a gantry girder and a special H-section is to be fabricated. The trial section is shown in Fig. 4.10. Check its suitability to support a factored axial load of 32 000 kN assuming both ends to be pinned over a length of 8 m. Steel of design strength 325 N/mm² is to be used. Could a rolled section be suitably reinforced (by welding cover plates to its flanges) so as to provide an alternative section?

Solution

$A = (60 \times 10) \times 2 + 50 \times 6 = 1500$ cm²
$I_y = 2(10 \times 60^3)/12 = 360 \times 10^3$ cm⁴ (neglects web)
$r_y = \sqrt{(360 \times 10^3/1500)} = 15.49$ cm
$\lambda = L/r_y = 8000/155 = 51.6$

From *Table 25*, noting that $t > 40$ mm, use curve *d*.
Since the section will be fabricated by welding and no guarantee that flame-cut plates are to be used is provided, use a reduced design strength

$p_y = 325 - 20 = 305$ N/mm².

Table 27d for $\lambda = 51.6$ and $p_y = 305$ N/mm² gives $p_c = 217$ N/mm²

$P_c = 228 \times 1500 \times 10^2 = 32\,550 \times 10^3$ N = __32 500 kN__, section is suitable

The heaviest rolled section is a 356 × 406 UC 634, the relevant properties for which are $A = 808$ cm², $I_y = 98\,211$ cm⁴ and $r_y = 11.0$ cm. Since this provides about one half of the area of the welded section it will need substantial cover plates. As a first trial use 400 mm × 100 mm plates on both flanges as shown in Fig. 4.11.

$A = 808 + 2(10 \times 40) = 1608$ cm²
$I_y = 98\,211 + 2(10 \times 40^3)/12 = 204\,878$ cm⁴
$r_y = 204\,878/1608 = 11.29$ cm
$\lambda = L/r_y = 70.9$

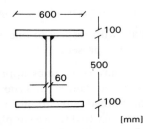

Fig. 4.10 Column cross-section of Example 4.5, welded section.

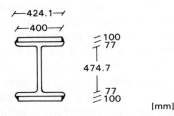

Fig. 4.11 Column cross-section of Example 4.5, reinforced rolled section.

From *Table 25*, noting section is as shown in *Table 26*, and $t > 40$ mm use curve *b*. The full p_y may be used.
From *Table 27b*, for $\lambda = 70.9$ and $p_y = 340$ N/mm^2, $p_c = 233$ N/mm^2
Hence compressive resistance $P_c = 233 \times 160\ 800 = 37\ 466 \times 10^3$ N
$$= \underline{37\ 466\ \text{kN}}$$

Since this exceeds the required resistance the section could be redesigned using smaller cover plates. It is left to the reader to show that 370×100 mm plates provide a compressive resistance of 34 830 kN for an area of 1548 cm^2 and a slenderness of 73.7.

This example clearly demonstrates the effect of making allowance for the variations in strength between different types of column. Although the areas of the two sections are similar, the reinforced UC is significantly more slender (λ of 70.9 compared with 51.6) and yet the compressive strengths of the two sections are very similar (233 N/mm^2 and 217 N/mm^2). The reason for this lies in the more favourable column curve assigned to the cover plated section as well as the use of a reduced design strength for the welded section.

The choice of section for a given application will depend on a number of factors, especially availability of materials and fabrication facilities, although it is worth noting that the reinforced section would occupy less space on plan.

4.3   INFLUENCE OF END CONDITIONS

In discussing the column curves of the previous section it was assumed throughout that both ends were supported such that:

(1) they could not translate with respect to one another
(2) no rotational restraint was present.

While conditions in practice may sometimes approximate to this, several other arrangements will also be encountered. True ultimate strength results for columns with other than pinned ends are not readily available. Even if they were it would still be necessary to devise a simplified treatment for use in design since the provision of a portfolio of column curves to cover all possible restraint conditions would be impractical. The usual approach for design consists of reducing the actual case under consideration to an equivalent pin-ended case by means of an effective-length factor determined from a comparison of elastic critical loads. This process therefore assumes that the influence of imperfections will be broadly similar for all forms of restraint, being a function of effective slenderness only.

The notion of an effective column length comes directly from elastic stability theory [3] where it is used as a device to relate the behaviour of columns provided with any form of support to the behaviour of the basic pin-ended case. Thus the general expression for critical load becomes

$$P_{cr} = \pi^2 EI/l^2 \tag{4.1}$$

where $\ell = kL$ is the effective length and k is termed the 'effective length factor'.

Taking, as an example, the case of a column with fixed ends, for which the critical load is

$$P_{cr} = 4\pi^2 EI/L^2$$

the effective length is obtained directly as

$$\ell = L/2, \text{or } k = 0.5 \tag{4.2}$$

Table 4.2 gives theoretical effective length factors for several standard cases. When used in the context of elastic critical loads the effective length also corresponds to the distance between points of inflection in the buckling mode [4]. An important general point to note from Table 4.2 is that when relative translation of the ends is prevented, k cannot exceed unity but that effective lengths up to several times the actual column height are possible for columns which are free to sway.

4.3.1 Design approach

Guidance on the choice of effective length factors for columns in simple construction is given in *Cl. 4.7.2* of BS 5950, particularly *Table 24*. Comparison with Table 4.2 shows the code values to be either equal to or slightly higher than the equivalent theoretical values. When higher values are specified it is usually in recognition of the practical difficulties of providing complete restraint against rotation. Further information on the appropriate effective column lengths to use in single storey and multistorey buildings of simple

Table 4.2
Effective length factor for columns

	Both ends pinned	Intermediate restraint	One end fixed	Both ends fixed	Cantilever
Support arrangements					
Value of k based on elastic critical load	1.0	0.5	0.7	0.5	2.0
BS 5950 design value of k	1.0	0.5	0.85	0.7	2.0

construction is provided in *Appendix D*. The decision as to what value is applicable to a particular case often requires considerable judgement; for situations in which the designer is uncertain of the degree of restraint present, the safe approach is always to neglect the restraint and to select a high value for k.

Example 4.5
Repeat Example 4.3 assuming that the column is built-in at its base and is supported at its top in such a way that deflection about the minor axis is prevented and deflection about the major axis is not.

Solution
Reference to *Table 24* shows that the appropriate effective lengths are

minor axis, $\ell_y = 0.85L$
major axis, $\ell_x = 2.0L$

Referring back to Example 4.3, $\lambda_y = 0.85 \times 3100/51.9 = 50.8$
$$\lambda_x = 2.0 \times 3100/89.8 = 69$$
Thus, because of the different degrees of restraint in the two planes, major-axis buckling is now more critical. From *Table 25* use curve b, hence using *Table 27b* for $p_y = 275$ N/mm^2 and $\lambda = 69$, value of $p_c = 204$ N/mm^2 and

$$P_c = 204 \times 7580 \text{ N} = \underline{1546 \text{ kN}}$$

Had the column also been restrained at its top about the major axis, then λ_x = 29.3 and minor-axis buckling would again have controlled, leading to P_c = 1652 kN. Comparing this with Example 4.3 shows that the change in restraint conditions (provision of rotational restraint at the base) produces an increase in strength of about 8%, at least the equivalent of a change in the column curve used.

4.4 SPECIAL TYPES OF STRUT

4.4.1 Angle sections

Single angles are often used as compression members in situations where comparatively low forces need to be transmitted, a common example being the roof truss shown earlier in Fig. 4.1a. In situations where a single angle could not provide sufficient compressive strength, or perhaps where the disparity in size between tension and compression members would make jointing difficult, double angles may be used. These are formed from two angles placed back to back, normally with a space between them to allow the joints at either end to be made via gusset plates in such a way that eccentricity of loading at the joint is minimized. It is, of course, necessary to ensure that the two sections function together as one compound member. Thus 'stitching' must be provided at sufficient intermediate points that the load for buckling of one angle between fasteners exceeds the load for overall buckling of the compound section. In this, as in any problem involving buckling of a single angle, it is important to remember that the weakest plane will be in the direction of the minor principal axis which does not of course, coincide with either rectangular axis.

4.4.1.1 Design approach

Rules for the design of angle struts are based largely on empirical data [9, 13, 14]idue to the difficulties associated with quantifying both end restraint conditions and the eccentricities of loading introduced by the joints. For continuous struts, i.e. where one length is 'run through' to form several members as might happen for example in the rafter of a roof truss, it is permissible to design the intermediate bays as axially loaded, with the effective length being taken as the actual length in that bay. For discontinuous struts (including the end bays of continuous struts) BS 5950 gives two procedures depending upon the type of end fixing. In the case of single angle struts (*Cl. 4.7.10.2*) these are:

(1) connection through one leg by two or more fasteners in line or the equivalent in welding, $\lambda = 0.7L/r_{aa} + 30 \not< 0.85L/r_{vv}$.

in which r_{aa} = radius of gyration about an axis through the centroid of the angle parallel to the gusset.

r_{vv} = the minimum radius of gyration

(2) single fastener or the equivalent in welding, $\lambda = 0.7L/r_{aa} + 30 \not< 1.0L/r_{vv}$ and in addition, $P_c \not> 0.8p_c A$.

Whereas the first of these includes some allowance for eccentricity of loading by using a pessimistic effective length, the second, because it would clearly be confusing to specify an effective length factor greater than 1.0 when end translation is prevented, allows for the (probably greater) effect of load eccentricity by assuming that part of the compressive resistance must be used to resist bending, a device that is similar to the use of effective area for tension members as described in *Section 3*. Similar rules are also given in *Cl. 4.7.10.3* for double angle struts. Because of the smaller eccentricities associated with this class of section these are less severe.

Example 4.6
Determine the compressive resistance of an $80 \times 80 \times 10$ equal-angle section in Gr. 43 steel when it is used as a strut over a length of 1.8 m. Assume a single fastener is provided at each end.

Solution
From section tables, $A = 15.1$ cm^2, $r_{min} = r_{vv} = 1.55$ cm
From *Cl. 4.7.10.2*, take $\lambda = 1.0 \times 1800/15.5 = 116$
From *Table 25*, use curve c
From *Table 27c*, for $\lambda = 116$ and $p_y = 275$ N/mm^2, value of $p_c = 102$ N/mm^2
From *Cl. 4.7.10.2*, $P_c = 0.8 \times 102 \times 1510$ N = 123 kN

If the end connections had been improved to two fasteners in line then P_c could be increased to 192 kN, an improvement of over 50%.

A summary of appropriate values of λ for angle, channel and tee-struts with various forms of eccentric connection, is provided in *Table 28*.

4.4.2　Laced and battened struts

The columns of industrial buildings are often called upon to provide support for a gantry crane. Quite heavy axial loads are therefore introduced into the lower portions of these columns. Rather than use a heavy section over the full height, a second member may be introduced over this lower length and the two legs connected together into the lattice arrangement shown as Fig. 4.12. Two slightly different forms may be used:

(1) the laced column in which relatively light transverse members are arranged in a triagulated fashion.
(2) the battened column in which rather heavier battens are placed only at right angles to the column axis.

4.4.2.1　Design approach

Design of laced and battened struts is similar in principle to the design of double angle struts in that the lacing or battens should be so arranged that they insure against premature local failure [15]. The strut may then be designed as a single integral member with a slenderness given by

$$\lambda = \sqrt{(\lambda_m{}^2 + \lambda_c)}$$

where　$\lambda_m = \ell/r$ for the whole member
　　　　$\lambda_c = \ell/r_{min}$ for the main component

subject to the limitations $\lambda \not> 50$ and $\lambda \not< 1.4\,\lambda_c$.

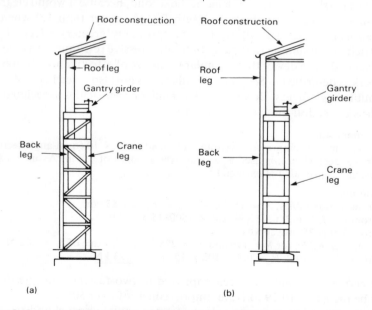

Fig. 4.12 Compound columns suitable for supporting crane gantries in industrial buildings: (a) laced column, (b) battened column. (*Bates, Constrado Publications.*)

Additional rules covering the proportioning of the transverse members and the arrangement of the fasteners are given in *Cl. 4.7.8, 4.7.9* and guidance on the assessment of effective lengths for intermediate portions of the main legs is given in *Appendix D*.

REFERENCES

1. Structural Stability Research Council. Technical Memorandum No. 3: Stub-column test procedure. Appendix B of *Guide to Stability Design Criteria for Metal Structures*, 3rd edn, edited by Johnston, B.G. Wiley-Interscience, New York (1976).
2. Dwight, J.B. and Moxham, K.E. Welded steel plates in compression. *Structural Engineer* **47** (2), 49–66 (February 1969).
3. Timoshenko, S.P. and Gere, J.M. *Theory of Elastic Stability* 2nd edn. McGraw-Hill, New York (1961).
4. Kirby, P.A. and Nethercot, D.A. *Design for Structural Stability*. Granada, St. Albans (1979).
5. Tall, L. Centrally compressed members. In *Axially Compressed Structures – Stability and Strength*, edited by Narayanan, R. Applied Science Publishers, London, 1–40 (1982).
6. Strymowicz, G. and Horsley, P. Strut behaviour of a new high yield stress structural steel. *Structural Engineer* **47** (2), 73–8 (February 1969).
7. Dwight, J.B. Strength in Compression. Revision of BS 449 *The Structural Use of Steelwork in Building Symposium*, Institution of Structural Engineers, London, 11–16 (1978).
8. Dwight, J.B. Adaptation of Perry formula to represent the new European steel column-curves. *Steel Construction*, AISC, **9** (1), (1975).
9. ECCS, Manual on the Stability of Steel Structures. Introductory Report, *Second International Colloquium on Stability*, Liège (1977).
10. Johnston, G.B., ed. *Guide to Stability Design Criteria for Metal Structures*, 3rd edn, Wiley-Interscience, New York (1976).
11. Chen, W.F. and Atsuta, T. *Theory of Beam-Columns*, vol. 1, McGraw-Hill, New York (1976).
12. Young, B.W. and Robinson, K.W. Buckling of axially loaded welded steel columns. *Structural Engineer* **53** (5), 203–7 (May 1975).
13. Kennedy, J.B. and Madugula, M.K.S. Buckling of single and compound angles. In *Axially Compressed Structures – Stability and Strength*, edited by Narayanan, R. Applied Science Publishers, London, 181–216 (1982).
14. Woolcock, S. and Kitipornchai, S. The design of single angle struts. *Steel Construction*, AISC. **14** (4), 2–23 (1980).
15. Porter, D. Battened columns – recent developments. In *Axially Compressed Structures – Stability and Strength*, edited by Narayanan, R. Applied Science Publishers, London, 249–78 (1982).

EXAMPLES FOR PRACTICE

✓ 1. Check whether a 406 × 140 UB 39 in Grade 43 steel would be affected by local buckling effects when used as a column.

[web limit exceeded, P_c = 919 kN]

2. Determine the axial load capacity of a short length of square box column in Gr. 50 steel fabricated by welding together four 800 × 20 mm plates.

[12 992 kN]

3. Determine the capacity of a 254 × 254 UC 107 in Grade 43 when used as an axially loaded column of effective length 4.2 m.

[2636 kN]

4. Select the lightest UC in Grade 43 steel that is capable of carrying an axial compressive load of 2100 kN.

[305 × 305 UC 118]

5. Determine the axial load capacity of a 90 × 90 × 8 mm angle section in Grade 43 steel when used as a column with an effective length of 1.2 m.

[257 kN]

6. Select the lightest equal leg angle in Grade 43 steel capable of carrying an axial compressive load of 295 kN over an effective height of 1.25 m.

[100 × 100 × 8 mm]

7. Determine the load-carrying capacity of a box section made from four 800 × 20 mm Grade 43 steel plates when used as an axially loaded column over an effective height of 10 m.

[10 656 kN]

8. Determine the compressive resistance of a 120 × 120 × 10 mm angle of Grade 43 steel when used as a strut over a length of 2.2 m, assuming:
   (a) Fastening with a single bolt at each end

[318 kN]

   (b) Fastening with two bolts in line at each end

[378 kN]

9. Select an unequal angle section in Grade 43 steel capable of sustaining an axial compressive load of 255 kN over a length of 1.8 m, assuming:
   (a) Fastening to a gusset through the longer leg with a single bolt

[150 × 90 × 10 mm]

   (b) Fastening to a gusset through the longer leg with at least two bolts in line

[125 × 75 × 10 mm]

5
Beams

One of the most frequently encountered types of structural member is the beam, the main function of which is to transfer load principally by means of flexural or bending action. In a typical rectangular building frame the beams would comprise the horizontal members which span between adjacent columns; secondary beams might also be used to transmit the floor loading into the main beams. For the more usual forms of structural framing it is normally sufficient to consider only bending effects, the influence of any torsional loading on the beams being relatively slight. Certain types of problem, such as design of crane girders do, however, require a proper allowance to be made for the effects of torsion.

The main forms of response for a beam subjected to simple uniaxial bending are listed in Table 5.1. Which of these will govern in a particular case depends principally upon the proportions of the beam, the form of the applied loading and the type of support provided. In addition to satisfying these strength limits it is also necessary to ensure that the beam does not deflect too much under the working loads, i.e. to satisfy the serviceability limit state.

5.1   IN-PLANE BENDING OF BEAMS OF COMPACT CROSS-SECTION

This discussion assumes that the beam's cross-section is such that the effects of local buckling may be neglected (a full discussion of this topic is presented in Section 5.3.1). The behaviour of a simple beam, which is constrained to deflect in the plane of the applied loading, under the action of a gradually increasing bending moment is illustrated in Fig. 5.1. Neglecting, for the present, the effect of residual stresses, the beam's response will be linear up to that value of the applied load W_y which just causes the maximum extreme fibre stress at the cross-section of greatest moment to reach the material yield strain ϵ_y. At higher loads, deformations will increase more rapidly until the

Table 5.1
Main failure modes for beams

Mode	Description	Illustration	Section	Comments
Excessive bending	Providing the beam is adequately braced in the lateral plane (stocky beam) and its component plate elements are not too thin (compact cross-section), then failure will take place by excessive deformation in the plane of the applied loading		5.1	Basic mode of failure if all others are prevented
Lateral torsional buckling	Failure occurs by a combination of lateral deflection and twist, the load at which this occurs being dependent upon the proportions of the beam, the way the loading is applied and the support conditions provided		5.2	Can be prevented by the provision of suitable lateral bracing
Local buckling	Failure occurs by buckling of a flange on compression or of the web due to shear or combined shear and bending or, where concentrated loads are applied, as a result of vertical compression		5.3 / 5.3.1	Unlikely for hot-rolled sections for which the proportions have been selected so as to minimize the importance of flange and web buckling; web stiffening sometimes required to prevent shear buckling in plate girders; bearing stiffeners sometimes required under point loads and at reaction points.

| Local failure | Several possibilities including:
(i) shear yield of web
(ii) local crushing of web
(iii) excessive curling of thin flanges
(iv) local failure around web openings (if present) |  5.3

5.4 | Likely only for short spans and/or deep beams; can be prevented by suitable web stiffening. Possible only for extreme sections with very wide flanges. Special provision may be required around large web holes, e.g. use of local reinforcement |

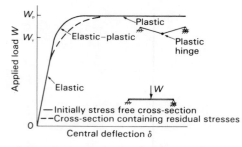

Fig. 5.1 Behaviour of simply supported steel beam. (*BSC Teaching Project, Imperial College, 1985.*)

fully plastic moment M_p is reached at the most highly stressed cross-section, whereupon a plastic hinge will form under the load. According to simple plastic theory [1, 2] deformations will now become uncontrolled. In practice the load-carrying capacity may actually be slightly greater due to the effects of strain hardening. However, it is customary to neglect this in design so that for simple beams, namely those that are not supported in such a way that redistribution of moment may occur, see Chapter 8, the formation of a plastic hinge at one point corresponds to the attainment of the ultimate load. The effect of the residual stresses which are normally present in structural sections is to cause yielding to start at a lower load with a consequent increase in the deflexions which occur at all subsequent load levels. However, the value of W_p is not affected because the residual strains must themselves be in equilibrium and cannot therefore alter the value of M_p.

Since the design for bending of laterally braced beams, i.e. those for which failure is governed by plastic action, is treated in BS 5950 as a special case of the more general problem involving consideration of lateral-torsional buckling, comments on the design approach will be delayed until section 5.2.

5.2 LATERAL-TORSIONAL BUCKLING OF BEAMS OF COMPACT CROSS-SECTION

5.2.1 Background to the problem

In much the same way that the design of all but the most stocky struts is controlled largely by considerations of overall instability so the design of most beams must be undertaken with a view to ensuring an adequate degree of safety against overall buckling. For beams the form of instability is, however, rather more complex since it involves both lateral deflection and twist as shown in Fig. 5.2. For the ideal case of a perfectly straight beam, loaded exactly in the plane of the web, theory [3–6] tells us that at the elastic critical load the beam will fail suddenly by deflecting sideways and twisting about its longitudinal axis; a form of response that may be observed in laboratory

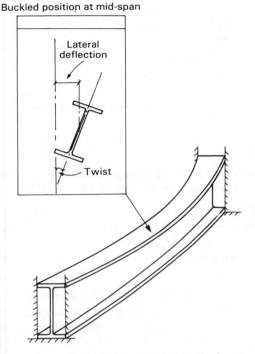

Fig. 5.2 Lateral–torsional buckling of a beam.

tests. Although the basic theory provides an adequate description of the behaviour of beams tested under very carefully controlled laboratory conditions, it does not cater for several of the factors which affect the lateral stability of beams in actual structures. Among the more important of these are initial bow and initial twist in the section, accidential eccentricites of loading and premature yielding due to the presence of residual stresses. Therefore, whilst elastic buckling theory assists in the identification of the governing parameters of the problem, proper use must also be made of representative test data if satisfactory design rules are to be established.

Experiments have demonstrated clearly that beams with closely spaced restraints can reach M_p while long unrestrained spans effectively fail by elastic lateral–torsional instability at moments that are very close to M_E, the theoretical elastic critical value [3–6]. Using the ratio of these two quantities as a measure of a beam's proneness to lateral–torsional collapse leads to the pictorial display of the problem shown in Fig. 5.3, where the quantity $(M_p/M_E)^{\frac{1}{2}}$ may be regarded as an 'effective slenderness for lateral–torsional buckling'. When test data are plotted on this basis it becomes possible to distinguish three regions of beam behaviour:

(1) Stocky beams: $(M_p/M_E)^{\frac{1}{2}} < 0.4$ for which M_p may be attained. (Beams for which plastic hinge action is possible are a subset of this requiring more closely specified limits; this topic is discussed in Chapter 8.)

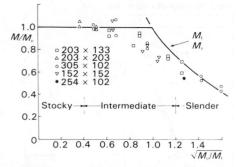

Fig. 5.3 Lateral–torsional buckling strength of steel beams of Gr. 55 steel. (*BSC Teaching Project, Imperial College, 1985.*)

(2) Beams of intermediate slenderness: $0.4 < (M_p/M_E)^{\frac{1}{2}} < 1.2$ which collapse through the combined effects of plasticity and instability at moments below either M_p or M_E.

(3) Slender beams: $(M_p/M_E)^{\frac{1}{2}} > 1.2$ which buckle at moments approaching M_E.

In the foregoing explanation it has simply been assumed that M_E corresponds to the theoretical elastic critical moment for the particular beam under consideration. Examination of the background theory [3–6] tells us that this quantity is a complex function of a number of parameters, the most important of which are the beam geometry, in particular its bending and torsional properties and its span, the type of restraint provided in the lateral plane and the pattern of moments (which will, of course, be affected by the conditions of support provided in the transverse plane). Thus the type of presentation of lateral buckling data used in Fig. 5.3 enables all of these factors to be conveniently accounted for.

5.2.2 Design approach

The basic design condition to ensure sufficient strength against overall buckling is given in *Cl. 4.3.7* of BS 5950 as

$$\overline{M} \not> M_b \tag{5.1}$$

in which $\overline{M}$ = equivalent uniform moment

$M_b = S_x p_b$ is the buckling resistance moment

and p_b = bending strength

S_x = plastic section modulus (for bending about the major axis)

Values of p_b for rolled-section beams are given in *Table 11a* in terms of the equivalent slenderness λ_{LT} which is defined as

$$\lambda_{LT} = \sqrt{\frac{\pi^2 E}{p_y}} \sqrt{\frac{M_p}{M_E}} \tag{5.2}$$

that is, the product of the quantity used as the abscissa in Fig. 5.3 and a constant for a given grade of steel. The limiting values of λ_{LT} for which p_b may be taken as p_y, leading to $M_b = M_p$, have been extracted from *Table 11* and are presented in Table 5.2 as λ_{L0}. Providing lateral bracing is employed at a spacing not exceeding λ_{L0}, no allowance for failure by lateral–torsional buckling is necessary.

The actual moment to be used in the design, M in eqn (5.1), may safely be taken as the maximum moment in the beam. Alternatively, for the arrangement of Fig. 5.4 in which the beam is loaded only at points of effective lateral restraint, producing an unrestrained length subjected only to unequal end moments, a reduced value may be used by following the rules of *Cl. 4.3.7.2* to obtain

$$\overline{M} = m\,M_{max} \tag{5.3}$$

in which $m = 0.57 + 0.33\beta + 0.10\,(\beta)^2 \not< 0.43$

     and β is the ratio M_1/M_2 of the moments at either end of the segment
        such that $1 \geqslant \beta \geqslant -1$

This special provision is based on the observation that results for moment gradient loading plot progressively higher on the frame of Fig. 5.3 as the ratio of M_1/M_2 decreases from 1.0 (single curvature) to -1.0 (double curvature). Thus for this form of loading only, λ_{LT} is always calculated on the basis of uniform moment ($\beta = 1.0$) and the allowance for the actual shape of the moment diagram is made by conducting the design check of eqn (5.1) using an 'equivalent uniform moment'. $\overline{M} = m\,M_{max}$.

Determination of the value of λ_{LT} is most conveniently undertaken by using the formula of *Cl. 4.3.7.5*, viz.

$$\lambda_{LT} = nuv\lambda \tag{5.4}$$

in which $\lambda = l/r_y$ is the minor axis slenderness
          $u = 0.9$ for rolled sections (see *Cl. 4.3.7.5*)

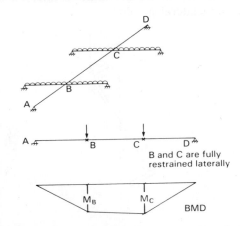

Fig. 5.4 Beam loaded at points of effective lateral restraint.

v = slenderness factor obtained from *Table 12*

n = slenderness correction factor (conservatively taken as unity but lower values may be used to take account of the pattern of moments as explained in *Cl. 4.3.7.6* which refers to *Tables 15* and *16* in which n-values are provided for several load cases), taken as unity if $m \neq 1.0$.

Table 5.2

Values of maximum slenderness λ_{LO} for which beam strength is not influenced by lateral–torsional instability and $p_b = p_y$

P_y (N/mm²)	245	265	275	325	340	355	415	430	450
λ_{LO}	37	35	34	32	31	30	29	28	28

In determining v, use is made of the 'torsional index' x; providing u is taken as 0.9, x may be approximated by the ratio of the overall depth to mean flange thickness D/T.

The procedure of eqn 5.4 is effectively a way of bypassing the explicit calculation of M_p and M_E as required by eqn (5.2) in order to produce a much shorter calculation. Although this sacrifices something in accuracy, the effect on the final design is normally likely to be insignificant. Where accurate calculations are required, i.e. 'exact' values of u and x are needed for eqn (5.4), *Appendix B* gives the full formulae; 'exact' values for standard rolled sections should be listed in the next set of section tables [7].

Allowance for end supports which provide some measure of rotational restraint in the buckling plane is treated in *Cl. 4.3.5* which gives a set of effective length factors to be used when calculating λ. A second set of effective length factors is provided in *Cl. 4.3.6* for dealing with cantilevers [6, 8]. In both cases the 'destabilizing' load case corresponds to the situation in which a vertical load is applied to the top flange in such a way that it is free to move sideways as the beam tends to buckle in a lateral–torsional manner. As Fig. 5.5 shows, such loads produce an additional torsional effect leading to a reduction in the beam's lateral stability.

Fig. 5.5 Torsion produced by tup flange destabilizing load.

Example 5.1
Determine the buckling resistance moment for a 254 × 146 × 31 UB in grade 43 steel assuming the beam to be laterally unsupported over a 3 m span.

Solution
From section tables, r_y = 3.19 cm, D/T = 29.1, S_x = 394.8 cm³

$\lambda = \ell/r_y$ = 3000/31.9 = 94.0

Taking $x = D/T$ = 29.1 gives λ/x = 94.0/29.1 = 3.23
From *Table 14*, noting that N = 0.5, v = 0.900
Taking u = 0.9 gives λ_{LT} = 0.9 × 0.9 × 94.0 = 76
From *Table 11a*, for p_y = 275 N/mm² and λ_{LT} = 76, value of p_b = 174 N/mm²

M_b = 174 × 394.8 × 10³ N/mm = <u>68.7 kN m</u>

Thus for this example, lateral buckling reduces the bending resistance by (275 − 174)/275 = 0.37, one third, or turning the problem around, the maximum laterally unbraced span for which the full bending resistance ($M_p = S_x \times p_y$) can be achieved is about 1.35 m (corresponding to a value of λ_{L0} = 35).

Example 5.2
Select a suitable UB section for the main beam of the structural arrangement shown in Fig. 5.6 assuming the use of grade 43 steel.

Solution
The bending moment diagram is shown in Fig. 5.6(b). Noting that the two cross-beams provide full lateral restraint at B and C the design will be governed either by segment BC or by segment CD.

BC β = 1194/1362 = 0.88 and from *Table 18*, m = 0.94

$\overline{M}$ = 0.94 × 1362 = 1280 kN m

Using the procedure of Example 5.1 the lightest section capable of carrying 1280 kN m over a 3.2 m laterally unsupported span is a 762 × 267 × 173 UB for which M_b = 1475 kN m.

CD β = 0/1362 = 0.0 and from *Table 18*, m = 0.57

$\overline{M}$ = 0.57 × 1362 = 776 kN m.

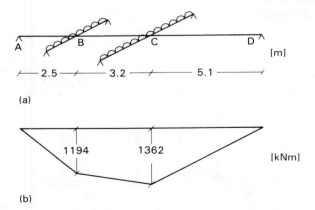

(a)

(b)

Fig. 5.6 Beam of Example 5.2: (a) loading and support conditions; (b) bending moment diagram.

Using the procedure of Example 5.1, for a moment of 776 kN m on a span of 5.1 m the 762 × 267 × 173 UB is safe, since M_b = 1072 kN m.
Thus the design is controlled by the lateral stability of segment BC and the chosen section is a 762 × 267 × 173 UB.

This example illustrates the use of the equivalent uniform moment concept when checking the strength of a beam that consists of several segments in the lateral plane. Often in such cases it is not possible to identify the critical segment simply by inspection. However, it is worth noting how, for this example, not using the equivalent uniform moment concept would require the provision of a section capable of carrying a moment of 1362 kN m over a span of 5.1 m, which would necessitate the use of a 914 × 305 × 201 UB with a corresponding increase in steel weight of 16%.

5.3 DESIGN OF BUILT-UP SECTIONS (PLATE GIRDERS)

For many structures all of the beams may be provided from among the standard range of rolled sections. However, from time to time situations will arise in which none of the available sections has sufficient capacity. Such problems occur normally when it is necessary to provide a long span and/or to support a particularly heavy load, one frequently encountered example being the gantry girders provided in industrial buildings to carry the rails for a large-capacity overhead travelling crane. The normal solution is to use a built-up section, commonly called a plate-girder, the proportions of which may be tailored specially to suit the design requirements. Nowadays it is normal practice to fabricate such sections simply by welding together three plates. However, in the past plate girders were often constructed by riveting or bolting, necessitating the use of angles to make the web to flange joints; several examples of this form of construction may still be seen. Different forms of plate girders are illustrated in Fig. 5.7.

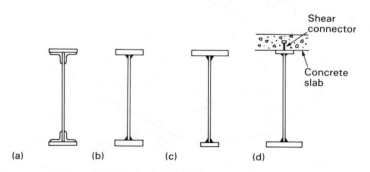

Fig. 5.7 Plate girder types: (a) with flange angles; (b) welded; (c) unequal flanges; (d) composite.

Because the designer has considerable freedom in proportioning a plate girder it is necessary for him to consider several structural problems which do not require the same attention when rolled sections are used. The most important of these are local buckling of the compression flange and shear buckling of the web [9]. Since the efficiency of the cross-section in resisting in-plane bending requires that the majority of the material be placed as far as possible from the neutral axis, it follows that minimum material consumption will frequently be associated with the use of a very thin web. However, if premature failure due to web buckling in shear is not to occur then web stiffening either by means of vertical stiffeners, horizontal stiffeners or a combination of the two will normally be required [9]. In practice, the choice between a thin web provided with stiffeners or a thicker web requiring no stiffening (and therefore involving lower fabrication costs) will depend upon a careful examination of the full costs of both forms of construction. Although flange capacity must also be checked, it is unusual for conventional plate girders to require compression stiffeners. On the other hand the ability of a slender web to resist both vertical buckling and/or local crushing will often prove to be inadequate without the assistance of suitable stiffening.

5.3.1 Local buckling effects in beams

The problem of local buckling in beams differs from that encountered in connection with columns (section 4.1.2) chiefly because of the greater variety of stress conditions present in the component plates of a beam. Even in the case of a compression flange, the design condition could vary from a requirement that strains approaching yield be accommodated, to one in which strains greatly in excess of yield must be accepted with no reduction in strength. In addition, the web will be subject to some combination of shear and bending due to the overall flexural action and possibly also to additional local stresses in the immediate vicinity of point loads. Thus it becomes necessary to check for each of the following forms of instability:

(1) Buckling of the compression flange, noting carefully the level of strain which the design moment implies.
(2) Buckling of the web in shear and/or bending.
(3) Vertical buckling of a portion of the web under concentrated loads or over reactions.

5.3.1.1 Flange local buckling

Figure 5.8 gives examples of the two classes of plate element identified by *Cl. 3.5.3* of BS 5950 as internal elements, which would correspond to the flange of a box beam, and outstand elements corresponding to the flange of the more commonly used I-section. For both types, four different ranges of 'compactness', each corresponding to a different performance requirement, are specified:

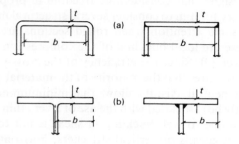

Fig. 5.8 Types of plate element: (a) internal elements; (b) outstand elements.
(*Dwight, Symposium on Revision of BS 449, 1978.*)

Class 1   Plastic ($b/t < \beta_1$). Able to attain yield with sufficient plastic plateau to permit the redistribution of moments within the structure required for plastic design.

Class 2   Compact ($\beta_1 < b/t < \beta_2$). Able to attain yield with sufficient plastic plateau to permit the section's full plastic moment to be attained.

Class 3   Semi-compact ($\beta_2 < b/t < \beta_3$). Able to attain yield but local buckling limits available plastic plateau so that the section's full plastic moment cannot be attained.

Class 4   Slender ($b/t > \beta_3$). Local buckling prevents the attainment of the material design strength.

A further distinction is made between welded and non-welded elements on account of the more severe effects of the locked-in residual stresses present in the former [10]. For Grades 43 and 50 steel the β limits of *Table 7* translate into the b/t limits given in Table 5.3. The moment capacity M_c of each of the four classes of section defined above is therefore calculated as:

(1) plastic $M_c = S p_y$
(2) compact $M_c = S p_y$
(3) semi-compact $M_c = Z p_y$ (5.5)
(4) non-compact $M_c < Z p_y$

Table 5.3
Limiting b/t values for plate elements subject to compression due to moment

		Internal element		Outstand element	
p_y values (N/mm^2)		275	355	275	355
Non-welded	β_1	26	23	8.5	7.5
	β_2	32	28	9.5	8.4
	β_3	39	34	15.0	13.2
Welded	β_1	23	20	7.5	6.6
	β_2	25	22	8.5	7.5
	β_3	28	25	13.0	11.4

where S and Z are the plastic and elastic section moduli respectively. Thus for non-compact sections the moment capacity must be reduced according to the geometrical proportions of the section. The simpler approach of *Cl. 3.6* in which a reduced value of p_y is used has previously been illustrated for columns in section 4.2. Consideration will now be given to the alternative. As the name suggests, this involves replacing the actual wide plate with a narrower 'effective width of plating' which is then assumed to be fully effective in compression.

The idea is well supported both by rigorous theory as well as by observations of the behaviour of compressed plating in tests. These show the relationship between effective width and actual width to be dependent principally upon the plate thinness b/t, the conditions of support along the longitudinal edges (internal or outstand element) and the severity of residual stress (welded or non-welded). Thus the moment capacity of a beam containing a non-compact compression flange must be calculated using the proportions of the effective cross-section as shown in Fig. 5.9. These correspond to the limits for semi-compact behaviour, i.e. any material in excess of the β_3 limit is ignored when calculating the section modulus Z.

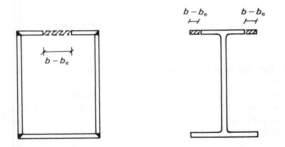

Fig. 5.9 Effective sections for determining the section modulus of members containing slender plate elements. (*Dwight, Symposium on Revision of BS 449, 1978.*)

Example 5.3
Check whether the moment capacity of a welded plate girder comprising two 650 × 25 mm flange plates and one 1500 × 15 mm web plate will be affected by flange local buckling, assuming (i) Grade 43 steel of design strength p_y = 265 N/mm², and (ii) Grade 50 steel of design strength p_y = 340 N/mm².

Solution
(i) For p_y = 265 N/mm², from *Table 7* maximum outstand b/t for flange to be compact = 8.5.
   Actual b/t, using *Fig. 3* = (325 − 15/2)/25 = 12.7, and $M_c < M_p$
   Maximum b/t for flange to be semi-compact = 13
   ∴ section is semi-compact and $M_c = Z p_y$
   I_x = (65 × 155³ − 63.5 × 150³)/12 = 2 311 614.6 cm⁴
   Z_x = 2 311 614.6/(75 + 2.5) = 29 827.3 cm³
   M_c = 265 × 29827 × 10³ = 7904 kN m
   S_x = 33 625 cm²

∴ reduction in capacity from that corresponding to compact behaviour

$$= \left(\frac{9247 - 7904}{9247} \right) = 14.5\%$$

(ii) For $p_y = 340$ N/mm², maximum b/t for flange to be semi-compact $= 11.7$.
∴ section is slender and assume b_e is limit for semi-compact behaviour
effective flange width $b_e = 11.7 \times 25 = 292$ mm
giving the effective section shown in Fig. 5.10.

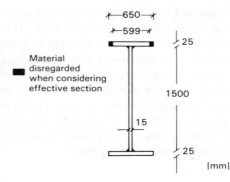

Fig. 5.10 Plate girder of Example 5.3.

Locate neutral axis by taking moments about the top edge as 793 mm from top edge.

I_x = $(599 \times 25)(793 - 12.5)^2 + (15 \times 1500^3/12) + (15 \times 1500) \, 18^2$
        $+ (650 \times 25)(757 - 12.5)^2$ $= 2.236 \times 10^{10}$ mm⁴

Z_x = (top flange) = $2.236 \times 10^{10}/793$ = 2.82×10^7 mm³

$M_c = 340 \times 282 \times 10^7$ = 9586 kN m

∴ reduction in capacity from that corresponding to semi-compact

behaviour $= \left(\dfrac{10\ 141 - 9586}{10\ 141} \right) = 5.5\%.$

Since the 'excess' material at the tips of the flanges cannot be included in calculations of the beam's moment capacity, consideration might be given to using a section which just meets the semi-compact requirements. This would avoid the complication of locating the neutral axis of the (effective) monosymmetric section. Using the simpler alternative of *Table 8* requires a 10% reduction in p_y.

5.3.1.2   Web behaviour

Girder webs will normally be subjected to some combination of shearing and bending stresses and, since the most severe condition in terms of web buckling is normally the pure shear case, it follows that it is those regions adjacent to supports or in the vicinity of point loads which generally control the design. Shear buckling occurs largely as a result of the compressive stresses acting diagonally within the web, as shown in Fig. 5.11, with the

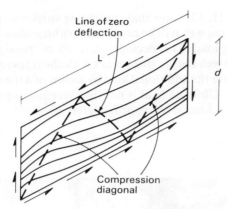

Fig. 5.11 Buckling of a girder web in shear. (*After ref. 5.*)

number of waves tending to increase with an increase in the panel aspect ratio L/d. The elastic critical stress q_e may be expressed as:

$$q_e = \left[0.75 + \frac{1}{(L/d)^2}\right]\left[\frac{1000}{d/t}\right]^2 \qquad \text{for } L/d \leqslant 1$$

$$q_e = \left[1 + \frac{0.75}{(L/d)^2}\right]\left[\frac{1000}{d/t}\right]^2 \qquad \text{for } L/d \geqslant 1$$

(5.6)

Because of the importance in eqn (5.6) of the plate aspect ratio, shear buckling resistance may conveniently be improved by dividing the web into a series of panels by using intermediate vertical stiffeners. Examination of eqn 5.6 suggests that a stiffener spacing which leads to panels having an aspect ratio a/d of between 0.5 and 2 will normally prove the most efficient. Although it is also possible to improve web strength by using horizontal stiffeners, this topic is not covered by BS 5950 (which simply refers the reader to the Bridge Code BS 5400) and is therefore beyond the scope of this text. Equations (5.6) form the basis of the design method for webs provided in *Cl. 4.4.5.3* of BS 5950, which gives the shear buckling resistance of a web as

$$V_{cr} = dtq_{cr} \qquad (5.7)$$

where q_{cr} = critical shear strength.

For stocky webs q_{cr} is simply the yield stress in shear, conveniently rounded to $0.6p_y$. The quantity used to define web slenderness λ_w is given by

$$\lambda_w = (0.6p_y/q_e)^{\frac{1}{2}} \qquad (5.8)$$

and stocky panels are regarded as those for which $\lambda_w < 0.8$. For slender panels, $\lambda_w > 1.25$, q_{cr} is taken as the elastic critical stress q_e and between the two limiting values of λ_w the following linear transition is employed:

$$q_{cr} = 0.6p_y[1 - 0.8(\lambda_w - 0.8)] \qquad (5.9)$$

For design purposes *Tables 21a–21d* provide values of q_{cr} directly in terms of p_y, a/d and d/t.

Experiments [9, 11, 12] show that, providing sufficiently heavy stiffeners are employed then the web will be capable of withstanding loads in excess of the elastic buckling load. This occurs as a result of 'tension field action' in which the diagonal web tensile stresses act with the transverse stiffeners and the flanges to transfer the additional load by means of a truss type of action as shown in Fig. 5.12. Ultimate load is not then reached until after the tension field has yielded at a load given approximately by

$$V_{\text{ult}} = V_{\text{cr}} + V_{\text{tf}} \tag{5.10}$$

(a)

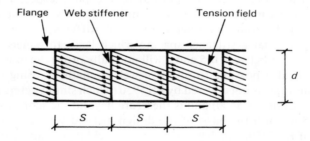

(b)

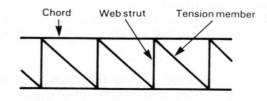

(c)

Fig. 5.12 Tension field action in plate girder webs. (a) Test girder showing well-developed tension fields (H.R. Evans); (b) load carrying mechanism of tension fields, web stiffeners and flanges; (c) equivalence to behaviour of a truss. (After ref. 5.)

in which V_{cr} is the elastic critical load and V_{tf} is the additional load due to tension field action.

BS 5950 permits the use of this 'basic tension field action' for all girders other than crane gantry girders, provided certain conditions are met. The most important of these is that the end panels are made sufficiently strong to anchor the longitudinal force set up by the tension field. Rules for the detailed design of end panels are given in *Cl. 4.4.5.4.2*. For all other panels the shear buckling resistance V_b may again be calculated using eqn (5.7), but with q_{cr} replaced by q_b, the basic tension field shear strength; values of q_b against p_y, d/t and a/d are tabulated in *Tables 22a–22d*.

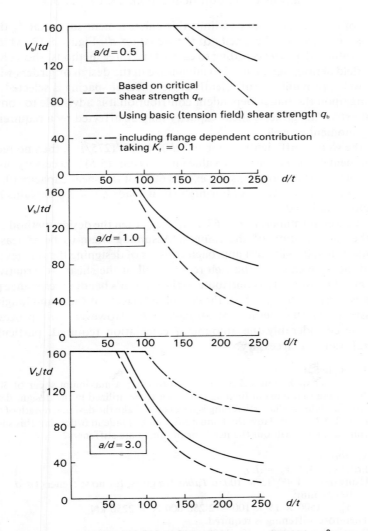

Fig. 5.13 Strength of web panels in shear, $p_y = 240$ N/mm².

One further refinement permitted by BS 5950 is the use of 'full tension field action', in which the additional contribution of sufficiently rigid flanges in anchoring the tension field is also taken into account, leading to a further increase in shear capacity. This permits V_b to be enhanced by a 'flange dependent contribution', leading to

$$V_b = (q_b + q_f \sqrt{K_f}) dt \tag{5.11}$$

in which q_b = basic (tension field) shear strength, *Table 22*

q_f = flange dependent shear strength factor, *Table 23*

K_f is a measure of the ability of the flanges to participate in the advanced tension field action, see *Cl. 4.4.5.4*.

It is, of course, necessary to limit this enhancement such that V_b does not exceed the shear yield capacity of the web $0.6p_y dt$. Figure 5.13, which compares values of V_b/td according to each of the three methods, shows how tension field action may be used to advantage in the design of girders with deep thin webs, providing a sufficiently close stiffener spacing is selected. Before relying upon the flange dependent contribution it is advisable to consider to what extent the girder's shear capacity may be affected by a requirement to carry moment as well.

If the web is sufficiently compact, $d/t < 63 \sqrt{(275/p_y)}$, then no reduction in moment capacity below the value given by eqn (5.5) is necessary providing the average shear force does not exceed 60% of the shear capacity $(0.6p_y dt)$. This will often be the case. For high shear loads *Cl. 4.2.6* explains how M_c should be reduced.

If the web is thinner, $d/t > 63 \sqrt{(275/p_y)}$, then the design method depends on the classification of the flanges. Assuming these to be at least semi-compact then the simplest approach consists of designing the flanges to withstand the moment with the web resisting all of the shear. Alternatively the web may be assumed to contribute to the section's bending resistance providing it is correctly designed for the combined effects of shear and longitudinal stresses using the method of *Appendix H*. However, this process does increase considerably the amount of calculation required, particularly if several trials are necessary.

Example 5.4
The girder of Example 5.3 is required to carry a maximum shear of 3000 kN. Assuming that tension field action is not to be utilized in the design, determine whether intermediate stiffening is necessary. Take the design strength of the steel p_y as 275 N/mm². How thick must the web be made in order that this same load can be carried without the need for intermediate stiffeners?

Solution
From eqn 5.7, $V_b = dt\, q_{cr}$
Using $d/t = 1500/15 = 100$ in *Table 21a* gives, for no stiffeners $(a/d = \infty)$, q_{cr} = 100 N/mm²
$\therefore V_b = 1500 \times 15 \times 100 = 2\,250\,000$ N = 2250 kN
Therefore stiffening is required.
Required $q_{cr} = 300 \times 10^3/1500 \times 15 = 133$ N/mm²

From *Table 21b*, for $d/t = 100$ max., a/d corresponding to this strength $= 1.15$
(round-off to 1.0).
∴ provide stiffeners at $1.15 \times 1500 = 1725$ mm intervals.
For the second part of this example a trial and error approach is necessary since
V_b depends on q_{cr} which is itself dependent on t. Clearly t must be greater than
15 mm.
Try $t = 18$ mm $\rightarrow d/t = 83.3$
and q_{cr} (for $a/d = \infty$) $= 127$ N/mm^2
∴ $V_b = 1500 \times 18 \times 127 = 3429$ kN

Because the web in this example is not particularly slender ($d/t \leqslant 100$) the
better solution is probably to increase its thickness and avoid the need for
stiffening. However, inspection of *Table 21b* shows that for deeper girders
comparatively much larger strength increases result from the use of
stiffeners, particularly closely spaced stiffeners. For example, for $d/t = 250$,
stiffeners at $0.9d$ double the shear strength while stiffeners at $0.4d$ produce at
least a six-fold improvement.

Example 5.5
Assuming a stiffener spacing equal to the panel depth, determine the shear
capacity of the girder of Example 5.3, assuming the use of tension field action.

Solution
From eqn 5.11 using only basic tension field action

$V_b = dtq_b$

Using $d/t = 100$ and a/d 1.0 in *Table 22b* gives $q_b = 151$ N/mm^2
∴ $V_b = 1500 \times 15 \times 151 = 3398$ kN

This is an increase of 50% on the value obtained using q_{cr}. Since *Table 23b*
gives $q_f = 173$ N/mm^2 reliance upon the flange contribution will increase this
up to the maximum (based on full shear yield) of 3713 kN even if K_f adopts
the extremely low value of 0.005.

Example 5.6
Select plate sizes for a welded plate girder of approximately 1.2 m depth, sufficient to withstand maximum coincident values of moment and shear of
570 kN m and 1200 kN assuming Grade 43 steel. The girder will be fully braced
against lateral instability.

Solution
Using *Cl. 4.4.4* the design basis will be to adopt semi-compact flanges, a thin web
stiffened as necessary and therefore to follow the method of *Cl. 4.4.4.2a* in using
the flanges to resist the moment and the web to resist the shear. Take $p_y =
265$ N/mm^2. (Assumes $t > 16$ mm).
For M to be resisted by the flanges, $A_f = 5700 \times 10^6/(265 \times 1200)$
$= 1792$ mm^2

Taking $B \approx D/2$ gives $B = 600$ mm
If flanges are to be semi-compact, *Table 7* gives $b/T \not> 13$
∴ $T \not< 300/14 = 23.1$ mm
Try 600 $\times$ 30 mm plates for flanges:
M_f 265 $\times$ (600 $\times$ 30) $\times$ (1200 + 30) = 5867 kN m
Try 8 mm web 1200 mm deep:
For d/t of 150, *Table 21b* gives $q_{cr} = 44$ N/mm^2
or *Table 22b* gives $q_b = 68$ N/mm^2

$\therefore$ using basic tension field action from eqn 5.11

$V_b = 68 \times 1200 \times 8 = 653$ kN

$\therefore$ either use web stiffeners or increase t

If using stiffeners and tension field action required value of

$q_b = 1200 \times 10^3/(1200 \times 8) = 125$ N/mm^2

From *Table 22b*, max. $a/d = 0.88$

Try vertical stiffeners at 1 m intervals to give

$a/d = 1.0/1.2 = 0.83$

From *Table 22b*, $q_b = 130$ N/mm^2

From eqn (5.11), $V_b = 130 \times (1200 \times 8) = \underline{1248 \text{ kN}}$

If tension field action cannot be used, *Table 21b* gives max. a/d or q_{cr} of 125 N/mm$^2 = 0.65$

$\therefore$ position stiffeners at 780 mm intervals

If an unstiffened web is preferred for q_{cr} of 125 N/mm^2, *Table 21b* gives max. $d/t = 85$

$\therefore$ use 1200×15 mm web plate.

Design of transverse stiffeners

Transverse stiffeners must be proportioned so as to satisfy two conditions:

(1) They must be sufficiently stiff not to deform appreciably as the web tends to buckle.

(2) They must be sufficiently strong to withstand the shear transmitted by the web.

Since it is quite common to use the same stiffeners for more than one task (for example the stiffeners provided to increase shear buckling capacity can also be used as load-bearing stiffeners to assist the web in carrying heavy point loads), the above conditions must also, in such cases, include the effects of any additional direct loading.

Condition (1) is covered by *Cl. 4.4.6.4* by requiring web stiffeners to have a second moment of area at least equal to:

$$I_s \not< 1.5d^3t^3/a^2 \qquad \text{for} \qquad a > d\sqrt{2}$$
$$I_s \not< 0.75dt^3 \qquad \text{for} \qquad a < d\sqrt{2} \tag{5.12}$$

these values being increased in accordance with *Cl. 4.4.6.5* when lateral forces and/or eccentrically applied transverse loads must also be carried by the stiffener. The strength requirement is checked by ensuring that the stiffener acting as a strut is capable of withstanding F_q, the difference between the shear actually present adjacent to the stiffener and the shear capacity of the (unstiffened) web, together with any coexisting reaction or moment. Since the portion of the web immediately adjacent to the stiffener will tend to act with it, this 'strut' is assumed to consist also of a length of web of $20t$ on either side of the stiffener centreline giving an effective section in the shape of a cruciform. Full details of this strength check are given in *Cl. 4.4.6.6*. If tension field action is being utilized then the stiffeners bounding the end panel must also be capable of accepting the additional forces associated with anchoring the tension field.

Example 5.7
Design a suitable vertical stiffener for the stiffened version of the girder of Example 5.4.

Solution
Since $a/d = 1.0$, use second expression in eqn (5.12) to give
$I_s \nleqslant 0.75 \times 1500 \times 15^3 = 380 \text{ cm}^4$
Assuming the use of double-sided stiffeners of (say) 15 mm plate, since

$$I_s = \frac{15(2b)^3}{12}$$

$b = [3 \times 3800000/30]^{\frac{1}{3}} = 73$ mm
∴ use a pair of 75×15 mm plates
Check strength using *Cl. 4.4.6.6*
$V = 3000$ kN $V_w = 2250$ kN
∴ $F_q = 3000 - 2250$ $= 750$ kN
No additional loads
Effective width of plate $= 20 \times 15 \times 300$ mm
$I_x = 578.4 \text{ cm}^4$ $A = 114.8 \text{ cm}^2$ $r_x = 22.45$ mm
Take effective length ℓ as $d = 1500$ mm (assumes no lateral restraint to flanges at stiffener position, *Cl. 4.5.1.4*).
∴ $\lambda = 1500/22.45 = 67$
From *Table 27c*, or $p_y = 275 \text{ N/mm}^2$ $p_c = 208 \text{ N/mm}^2$
∴ $P_q = 208 \times 11475 \text{ N} = 2387$ kN

Since this exceeds F_q, stiffener has adequate strength.

5.3.1.3   *Web buckling due to vertical loads*

The application of heavy concentrated loads to a girder will produce a region of very high stress in the part of the web directly under the load. One possible effect of this is to cause outwards buckling of this region rather as if it were a vertical strut with its ends restrained by the beam's flanges. This situation also exists at the supports where the 'load' is now the reaction and the problem is effectively turned upside down. It is usual to interpose a plate between the point load and the beam flange, whereas in the case of reactions acting through a flange this normally implies the presence of a seating cleat. In both cases, therefore, the load is actually spread out over a finite area by the time it passes into the web as shown in Fig. 5.14. This is referred to as 'dis-

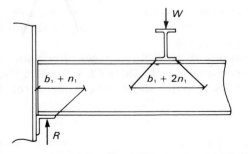

Fig. 5.14 Dispersion of concentrated loads and reactions.

persion into the web' and is controlled largely by the dimensions of the plate used to transfer the load, which it itself termed 'the stiff length of bearing'.

Because it is virtually impossible to provide anything approaching a rigorous theoretical treatment of this problem, design methods are based normally upon empirical formulae derived directly from tests. Thus *Cl. 4.5.2.1* of BS 5950 assumes the load to be carried by a vertical strut, the width of which is dependent upon the stiff length of bearing provided. This leads to the following expression for web buckling strength:

$$P_w = (b_1 + n_1)\,tp_c \tag{5.13}$$

in which b_1 = stiff length of bearing given by *Cl. 4.5.1.2*
n_1 = length obtained by dispersion of 45° through half the depth of the section
t = web thickness
p_c = compressive strength according to curve *c*

It is usual to assume that both flanges provide full rotational restraint to this 'strut' in which case λ may be taken as 2.5 *d/t* corresponding to a strut effective length of 0.7*d*. However, in situations where movement of one flange relative to the other is possible larger slendernesses are appropriate, as explained in *Cl. 4.5.1.4*.

It will often be the case that an otherwise satisfactory girder will prove to have inadequate strength according to eqn (5.13). One remedy is to employ load-bearing stiffeners to carry the excess load. Indeed this problem is encountered so frequently that designers will often call for such stiffeners at load and reaction points as a matter of course. Moreover it is not confined to built-up girders; many UB sections have webs that will be found to be inadequate when checked against eqn (5.13).

The design of load-bearing stiffeners is essentially the same as the design of vertical stiffeners for strength, as explained in the previous section. The load is again assumed to be resisted by a strut comprising the actual stiffeners plus a length of web of 20*t* on either side, giving an effective cruciform section. Providing the loaded flange is laterally restrained the effective length of this 'strut' may be taken as 0.7*L*. Although no separate stiffness check is necessary, load-bearing stiffeners must be of sufficient size that were the full load to be applied to them acting independently, i.e. on a cross-section consisting of just the stiffeners, then the stress induced should not exceed the design strength by more than 25%.

The exact functions of the different types of web stiffener that might be required on a slender web is explained in *Cl. 4.5*, which also refers the reader to the sections of BS 5950 that should be considered for the design of each type.

Example 5.8
For the girder of Example 5.3 check whether a 3000 kN reaction can be carried, assuming it to act through a cleat of 15 mm thickness.

Solution

From eqn 5.13 $P_w = (b_1 + n_1) t p_c$

From *Cl. 4.5.1.2* $b_1 = 2 \times (15 + 25) = 80$ mm

From *Cl. 4.5.2.1* $n_1 = d/2 = 750$ mm

From *Cl. 4.5.2.1* $\lambda = 2.5 \times 1500/15 = 250$

Using *Table 27c* $p_c = 28$ N/mm^2

$\therefore \ P_w = (80 + 750) \, 15 \times 28 \times 10^{-3} = 349$ kN

and web stiffeners are required.

Assuming $p_c = 200$ N/mm^2 (λ is likely to be low), required area of strut comprising stiffener + attached plating = $300 \times 10^3/200 = 15\,000$ mm^2

If using double-sided stiffeners of 20 mm plate, stiffener width needs to be
$\frac{1}{2}(1500 - (2 \times 300 \times 15))/20 = 150$ mm

$\therefore$ try 150 $\times$ 20 mm stiffeners

$I_x = 20 \times 315^3/12 \qquad \qquad = 52.1 \times 10^6$ mm^4

$A = 2(20 \times 150) + 600 \times 15 = 15\,000$ mm^2

$r_x = 58.9$ mm

From *Cl. 4.5.1.4* $L_E = 0.7d$

$\therefore \ \lambda = 0.7 \times 1500/58.9 = 17.8$

and from *Table 27c* $p_c = 263$ N/mm^2

$\therefore \ P_q = 15\,000 \times 263$ N = $\underline{3945 \text{ kN}}$

REFERENCES

1. Neal, B.G. *Plastic Methods of Structural Analysis*. Chapman & Hall, London (1970).
2. Horne, M.R. *Plastic Theory of Structures* 2nd edn. Pergamon, Oxford (1979).
3. Kirby, P.A. and Nethercot, D.A. *Design for Structural Stability*. Granada, St Albans (1979).
4. Timoshenko, S.P. and Gere, J.M. *Theory of Elastic Stability* 2nd edn. McGraw-Hill, New York (1961).
5. Trahair, N.S. *The Behaviour and Design of Steel Structures*. Chapman & Hall, London (1977).
6. Nethercot, D.A. Elastic lateral torsional buckling. In *Stability and Strength of Beams and Beam-Columns* edited by Narayanan, R. Applied Science Publishers, London (1983), 1–34.
7. BCSA/Constrado. *Structural Steelwork Handbook*. British Constructional Steelwork Association and Constrado, London (1978).
8. Nethercot, D.A. The effective lengths of cantilevers as governed by lateral buckling. *The Structural Engineer* **51** (5), 161–8 (1973).
9. Evans, H.R. Longitudinally and transversely reinforced plate girders. In *Stability and Strength of Plated Structures* edited by Narayanan, R. Applied Science Publishers, London (1984), 1–38.
10. Dwight, J.B. and Moxham, K.E. Welded steel plates in compression. *The Structural Engineer* **47** (2), 49–66 (February 1969).
11. Rockey, K.C., Evans, H.R. and Porter, D.M. The design of stiffened web plates – a state of art report. In *The Design of Steel Bridges* edited by Rockey, K.C. and Evans, H.R. Granada, London (1981), 215–42.
12. Rockey, K.C. and Skaloud, M. The ultimate load behaviour of plate girders loaded in shear. *The Structural Engineer* **50** (1), 29–47 (January 1972).

EXAMPLES FOR PRACTICE

✓ 1. Select a UB section capable of safely carrying a total uniformly distributed load of 170 kN over a span of 7.2 m, assuming the use of Grade 43 steel and the provision of full lateral support to the beam.

[305 × 165 UB 40]

✓ 2. Determine the buckling resistance moment for a 457 × 152 UB 60 in Grade 43 steel when it is simply supported over a span of 3.5 m.

[190 kN m]

✓ 3. Determine the buckling resistance moment for a 356 × 127 UB33 in Grade 43 steel for a span of 4.2 m, assuming that the applied loading produces moments which vary linearly from a maximum at one end to one quarter of this value at the other, both values being in a clockwise sense.

[92.5 kN m]

✓ 4. Select a UB in Grade 43 steel capable of safely carrying end moments of 640 kN m and 128 kN m over a laterally unsupported span of 6.5 m assuming that the moments produce single curvature bending.

[610 × 229 UB 125]

5. What is the moment capacity of a short length of welded plate girder fabricated from two 600 × 30 mm flange plates and one 1600 × 12 mm web plate assuming Grade 43 steel? What changes, if any, are required in plate thicknesses if the section is to be capable of carrying its full plastic moment?

[8943 kN m, T = 35 mm, t = 18 mm]

6. Determine the buckling resistance moment for a welded plate girder comprising 500 × 25 mm flange plates and a 1200 × 12 mm web plate in Grade 43 steel assuming a laterally unbraced span of 6 m

[4022 kN m]

7. A plate girder web is to be fabricated from a plate 1300 mm deep by 12 mm thick. Assuming Grade 43 steel, determine at what spacing vertical stiffeners must be placed if the girder is to be capable of carrying a shear load of 1350 kN without the use of tension field action. What would be the percentage increase in load carrying capacity if basic tension field action were permitted?

[1.82 m spacing, 53%]

8. Using the method of Cl. 4.4.4, design a plate girder in Grade 43 steel of approximately 1250 mm overall depth to withstand coincident moment and shear loads of 700 kN m and 2000 kN. Indicate the spacing of vertical stiffening, if any, necessary.

[Many solutions are possible but 700 × 35 mm flanges and a 1200 × 12 mm Web with stiffeners at 1450 mm spacing would be satisfactory]

6

Members under combined axial load and moment

Chapter 3–5 have dealt with the design of members subjected to a single form of loading, such as tension, and bending about one axis. However, situations will often arise in which the loading on a member cannot reasonably be represented as a single dominant effect. Such problems require an understanding of the way in which the various structural actions interact with one another. In the simplest cases this may amount to nothing more than a direct summation of load effects. Alternatively for more complex problems, careful consideration of the complicated interplay between both the individual load components and the resulting deformations is necessary.

The design approach discussed in this chapter is intended for use in situations where a single member is to be designed for a known set of end moments and forces. As such it is applicable to members in 'simple construction' although, as will be explained in Chapter 8, similar approaches are also possible for certain framing arrangements which fall within the general classification of 'continuous construction'.

Because of the additional complexity due to buckling associated with compressive loads, it is convenient to deal with the cases of tension plus bending and compression plus bending separately.

6.1 COMBINED TENSION AND MOMENTS

The procedures outlined previously in Chapter 3 for angle ties are valid only for those cases in which bending is produced solely by the fairly small eccentricities between the loaded leg and the member axis. For more general problems each load component must be considered separately since it is not known in advance which will be dominant.

The assumption of elastic behaviour leads to a simple design approach based on limiting the sum of the individual stresses at a cross-section to the design strength of the material p_y.

$$p_a + p_{bx} + p_{by} \not> p_y \tag{6.1}$$

in which p_a = axial stress due to load F

p_{bx} = maximum bending stress due to moments M_x about the x–x axis

p_{by} = maximum bending stress due to moments M_y about the y–y axis

Converting this to an expression for loads and rearranging, gives

$$\frac{F}{A p_y} + \frac{M_x}{Z_x p_y} + \frac{M_y}{Z_y p_y} \not> 1 \tag{6.2}$$

in which Z_x = elastic section modulus about the x–x axis

Z_y = elastic section modulus about the y–y axis

It has already been explained in Chapter 5 how stocky beams of compact cross-section may be expected to develop their full plastic moment capacity $M_p = S p_y$. Therefore, in order that eqn (6.2) reduces in the limiting cases of $F \to 0$ to the design condition for beams, the quantities $Z_x p_y$ and $Z_y p_y$ should be replaced by M_{cx} and M_{cy}, the cross-sectional moment capacities obtained from eqn (5.5) as explained in Chapter 5, to give

$$\frac{F}{A_e p_y} + \frac{M_x}{M_{cx}} + \frac{M_y}{M_{cy}} \not> 1 \tag{6.3}$$

in which A_e is the effective area, see Chapter 3.

Use of the major-axis bending cross-sectional strength in eqn (6.3) means that no allowance is made for lateral–torsional buckling effects, i.e. by using M_b from eqn (5.1) for M_{cx}. Although the presence of an axial tension may be expected to reduce any tendancy towards instability, it would seem prudent in cases where F is small and M_b is significantly less than M_{cx} not to disregard this since, in the limiting case of $M_y = 0$ and $F \to 0$, eqn (6.3) should agree with eqn (5.1). In the absence of clear evidence to the contrary it is suggested that the same allowance be made for all values of axial tension and eqn (6.3) be checked using M_b for M_{cx}; this may well be rather conservative in many instances.

Clause 4.8.2 of BS 5950 uses eqn (6.3) to check members at the points of maximum tension and bending; it suggests that this will usually be the ends. This is a linear interaction in which each of the three terms have equal effect. Figure 6.1 shows how it correctly tends towards the previously derived design conditions for the component cases as one form of loading becomes dominant.

More sophisticated analysis of this problem using the principles of plastic theory [1, 2] has shown that for compact cross-sections, i.e. those satisfying the geometrical limits of *Table 7* for no reductions in strength due to local buckling effects, eqn (6.3) may be replaced by

$$\left(\frac{M_x}{M_{rx}}\right)^{z_1} + \left(\frac{M_y}{M_{ry}}\right)^{z_2} \not> 1 \tag{6.4}$$

in which M_{rx} = reduced moment capacity about the x–x axis in the presence of the axial load F

$\quad\quad\;\; M_{ry}$ = reduced moment capacity about the y–y axis in the presence of the axial load F

$\quad\quad\;\; z_1$ = 2.0 for I- and H-sections, 5/3 for solid and closed hollow sections and 1.0 in all other cases

$\quad\quad\;\; z_2$ = 1.0 for all sections other than solid and closed hollow sections for which a value of 5/3 may be used

Use of eqn (6.4) will normally lead to higher results, as shown in Fig. 6.1. In using eqn (6.4) the values of M_{rx} and M_{ry} for standard sections may be obtained from section tables [3]. Alternatively the following expressions may be used for rolled I- and H-sections.

$$\begin{aligned}
S_{rx} &= (1 - 2.5n^2)\, S_x && \text{for } n < 0.2 \\
S_{rx} &= 1.125\,(1 - n)\, S_x && \text{for } n > 0.2 \\
S_{ry} &= (1 - 0.5n^2)\, S_y && \text{for } n < 0.447 \\
S_{ry} &= 1.125\,(1 - n^2)\, S_y && \text{for } n > 0.447
\end{aligned} \tag{6.5}$$

in which S_{rx}, S_{ry} = reduced plastic modulus in the presence of axial load F

$\quad\quad\;\; S_x, S_y$ = plastic modulus for zero axial load

$\quad\quad\;\; n$ = F/Ap_y

Values of plastic section moduli for angles bent about their rectangular axes are available [4]; for other types of cross-section, for example channels and fabricated I-sections, it is necessary to refer to texts on plasticity theory [1, 5]. In order that eqn (6.4) be consistent with the procedures of section 5 for

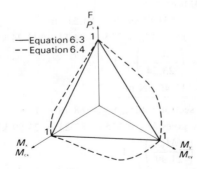

Fig. 6.1 Interaction for strength under combined loading.

simple bending, the values of M_{rx} and M_{ry} used should not exceed $1.2p_yZ_x$ and $1.2p_yZ_y$ respectively.

Example 6.1
Check whether a stocky $254 \times 146 \times 31$UB of grade 43 steel is safe under (factored) loads $F = 340$ kN and $M_x = 85.0$ kN m.

Solution
Since the member is compact take $M_{cx} = p_bS_x$

From eqn (6.3), $\dfrac{340 \times 10^3}{3990 \times 275} + \dfrac{85 \times 10^6}{394.8 \times 10^3 \times 275} = 0.310 + 0.783$

$$= \underline{1.093}, \text{ and section is not safe.}$$

Using eqns (6.4, 6.5), $n = \dfrac{340 \times 10^3}{3990 \times 275} = 0.310$

$\therefore\ S_{rx} = 1.125\ (1 - 0.310)\ 394.8 = 306.5\ \text{cm}^3$

Hence $\dfrac{M_x}{M_{cr}} = \dfrac{85.0 \times 10^3}{305.1 \times 275} = \underline{1.008}$, and section is not safe.

Using eqn (6.4) and section tables [3]

$S_{rx} = 394.8 - 653.9n^2 \qquad \text{for } n < 0.357$
$\quad\ = 394.8 - 653.9 \times 0.310^2 = 332.0\ \text{cm}^2$

Hence $\dfrac{M_x}{M_{rx}} = \dfrac{85.0 \times 10^3}{330.7 \times 275} = \underline{0.931}$, and section is safe.

This example shows how the use of progressively more 'exact' procedures leads to corresponding increases in the predicted capacity.

Example 6.2
If M_x is reduced to 60.0 kN m what moment may safely be applied about the minor axis.

Solution
From section tables, $S_y = 88.78\ \text{cm}^3$

From eqn (6.3), $\dfrac{340 \times 10^3}{3990 \times 275} + \dfrac{60 \times 10^6}{394.8 \times 10^3 \times 275} + \dfrac{M_y}{88.78 \times 10^3 \times 275}$

$$= 1.0$$

gives $\underline{M_y = 3.36\ \text{kN m}}$

Using eqns (6.4, 6.5) noting that $z_1 = 2$ and $z_2 = 1$

$S_{ry} = (1 - 0.5 \times 0.310^2)\ 88.78 = 84.51\ \text{cm}^2$

gives $M_{ry} = 275 \times 84.51 = 23.24\ \text{kN m}$

$\therefore\ \left(\dfrac{60}{84.28}\right)^2 + \left(\dfrac{M_y}{23.24}\right) = 1$

$\underline{M_y = 11.46\ \text{kN m}}$

Using eqn (6.4) and section tables, $S_{ry} = 88.78 - 15.86 \times 0.313^2 = 87.23\ \text{cm}^3$

gives $M_{ry} = 275 \times 87.23 = 23.99\ \text{kN m}$

$\therefore\ \left(\dfrac{60}{91.3}\right)^2 + \left(\dfrac{M_y}{23.99}\right) = 1$

$\underline{M_y = 13.63\ \text{kN m}}$

Once again eqn (6.4) gives a significantly higher result than eqn (6.3), with the use of the larger M_r values obtained from section tables producing a further improvement.

6.2 COMBINED COMPRESSION AND MOMENTS

When the axial component of the loading is compressive then the member's strength may be limited by either of the two conditions:

(1) Local capacity at the most heavily loaded cross-section.
(2) Overall buckling.

The first of these is essentially equivalent to the problem discussed above, while the overall buckling of a beam-column closely resembles column stability as discussed in Chapter 4. However, because the loading may take several different forms, so the member's response must be treated under a number of different headings.

The most common form of beam-column problem in building structures is the vertical member supporting (usually horizontal) beams; a typical example is shown in Fig. 6.2. Because of the assumptions regarding connection behaviour associated with 'simple construction', the loading on the stanchion may be taken as that shown in Fig. 6.3, i.e. an axial load F due to accumulated load from the floors above plus moments due to the beam reactions F_x and F_y assumed to act at known eccentricities e_y and e_x to the column faces. Guidance on the choice of suitable values for these eccentricities is provided in *Cl. 4.7.6* of BS 5950. Thus, in the most general case, the beam-

Fig. 6.2 Typical arrangement of beams and columns in a multistorey building.

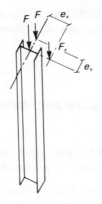

Fig. 6.3 Loading on a beam column in 'simply designed' frame.

column is subject to compression plus moments about both axes. If the loading and/or the beam arrangement is different at different levels then these moments will not be the same at both ends, that is moment gradients will exist as shown in Fig. 6.4. Of course, if some beams are absent or in the case where similar beams on opposite sides of the column carry identical loads so that the beam moments exactly balance, then the loading may reduce to a simpler form. Three distinct cases may be identified as shown in Fig. 6.5:

(1) The thrust is applied with an eccentricity about the minor axis (or if the eccentricity is about the major axis then either the column is prevented from deflecting out of this plane, by properly designed cladding for example, or there is no tendancy for out-of-plane buckling due to the

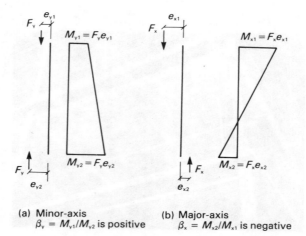

(a) Minor-axis
$\beta_y = M_{y1}/M_{y2}$ is positive

(b) Major-axis
$\beta_x = M_{x2}/M_{x1}$ is negative

Fig. 6.4 Bending moments in a beam column. (a) Minor axis $(\beta_y = M_{y1}/M_{y2}$ is positive); (b) major axis $(\beta_x = M_{x2}/M_{x1}$ is negative).

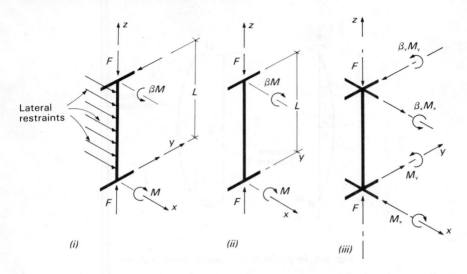

Fig. 6.5 Three classes of beam column problem: (a) In-plane behaviour: column deflects v in yz plane only [$F + M_x$ with bracing; $F + M_y$]; (b) flexural–torsional buckling: column deflects v in yz plane, then buckles by deflecting u in xz plane and twisting ϕ [$F + M_x$]; (c) biaxial bending: column deflects u, v and twists ϕ [$F + M_x + M_y$].

applied moment, as happens when the member is a circular tube) in which case the member will collapse by excessive deformation in this plane.

(2) The thrust is applied with an eccentricity about the major axis and the member fails by a combination of bending about the weak axis and twisting, similar to lateral–torsional beam buckling.

(3) The thrust is applied with an eccentricity about both axes in which case the member will collapse by combined bending and twisting.

Thus case (1) represents an interaction between column buckling and simple uniaxial beam bending, case (2) represents an interaction between column buckling and beam buckling, and case (3) represents the interaction of column buckling and biaxial beam bending. Clearly case (3) is the most general case with the others being more limited versions.

Not surprisingly the analytical background to the beam-column problem is extremely complex. The next three sections therefore provide only a relatively simple description; readers who are interested in obtaining a more complete understanding are advised to consult references [2, 6, 7].

6.2.1 Case (1): In-plane strength

Within the elastic range, case (1) of Fig. 6.5 may be analysed using the basic Euler theory of compression members [2]. Assuming equal end moments M

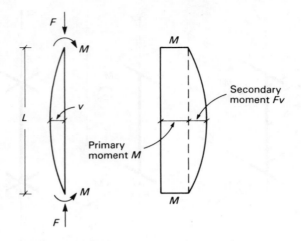

Fig. 6.6 In-plane behaviour of beam column.

as shown in Fig. 6.6 and setting up and solving the resulting differential equation, permits the deflected form and hence the bending moments and stresses in the beam-column to be determined. Because of the additional bending deformations caused by the compression F acting through an ever-increasing effective eccentricity (the lateral deformations v), the member will respond in a nonlinear fashion to the applied loads as shown in Fig. 6.7. The theoretical upper limit of F will be the elastic critical value $P_{cr} = \pi^2 EI/L^2$. However, this assumes indefinite elastic behaviour. If the stress due to compression

$$f_a = \frac{F}{A} \tag{6.6}$$

together with the maximum bending stress

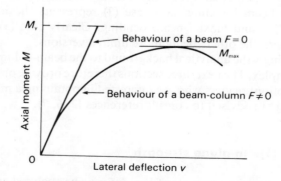

Fig. 6.7 Non-linear response of a beam column assuming elastic behaviour.

$$f_b = \frac{M_{max}}{Z} \tag{6.7}$$

is limited to the material yield stress σ_y, noting also that as $M \to 0$ F must be limited to P_{cr}, then the corresponding values of F and M will be related by

$$\frac{F}{P_{cr}} + \frac{M}{M_y} \left(\frac{M_{max}}{M} \right) = 1.0 \tag{6.8}$$

in which (M_{max}/M) allows for the additional secondary moments due to deformation. Rather than use the exact expression for (M_{max}/M), it is convenient to replace this with the close approximation

$$(M_{max}/M) \simeq \frac{M}{M_y (1 - F/P_{cr})} \tag{6.9}$$

This quantity is often termed an 'amplification factor' since it amplifies the primary moment M to give the total moment (primary + secondary). Thus eqn (6.8) becomes

$$\frac{F}{P_{cr}} + \frac{M}{M_y (1 - F/P_{cr})} = 1.0 \tag{6.10}$$

At low slendernesses, when P_{cr} will be so large that the amplification factor will have negligible effect, it plots as a straight line interaction between the axial (F/P_{cr}) and bending (M/M_y) effects. However, as slenderness increases so the effects of secondary bending become more significant, resulting in an increasingly concave interaction as shown in Fig. 6.8.

More rigorous analysis of this problem [2] allowing for the effects of yielding, residual stress, initial lack of straightness, etc., namely all these factors present in the behaviour of real steel members as discussed in Chapter 4, shows that the actual strength of beam-columns may be quite closely

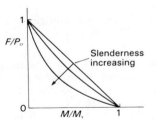

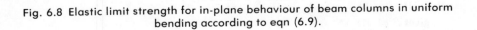

Fig. 6.8 Elastic limit strength for in-plane behaviour of beam columns in uniform bending according to eqn (6.9).

predicted using a modified version of eqn (6.10). Thus *Cl. 4.8.3.3.2* of BS 5950 uses

$$\frac{F}{P_c} + \frac{1}{2}\frac{F}{P_c}\frac{M}{M_c} + \frac{M}{M_c} = 1 \tag{6.11}$$

in which P_c and M_c are the uniaxial compressive and bending strengths and the product term approximates the function of the amplification factor.

6.2.1.1 Effect of non-uniform moment

Returning to elastic analysis and Fig. 6.6, if the end moments are now taken as M and βM, where $1 \geqslant \beta \geqslant -1$ and $\beta = 1$ corresponds to uniform single curvature bending, it may be shown [2] that the first yield interaction adopts the form of Fig. 6.9. As the applied moments tend towards double curvature ($\beta \to -1$), so the primary and secondary bending effects become less directly additive, with the result that the interaction plots higher. Eventually a situation will be reached in which yield occurs first at one end under the action of the primary moment alone, corresponding to the intersection with the zero slenderness (strength) interaction boundary. It is possible to represent these results quite accurately using eqn (6.9), providing an equivalent value $\overline{M} = mM$ is used. Coincidentally the relationship between m and β is very similar to that introduced in Chapter 5 for dealing with the lateral–torsional buckling of beams

$$m = 0.57 + 0.33\beta + 0.10\beta^3 \nleqslant 0.43 \tag{6.12}$$

Since eqn (6.10) now checks overall stability of the member, it is, of course, also necessary to ensure against local overstressing at the more heavily loaded end using the full value of the moment, i.e. to keep within the upper boundary of Fig. 6.9.

Based on the results of rigorous theoretical studies [2] together with test data it has been found that the 'equivalent uniform moment' concept may be used with the design expression of eqn (6.11). Thus M may now be reinterpreted as $\overline{M}$. In such cases it is necessary to check local strength separately using eqn (6.3) or eqn (6.4).

Example 6.3
What is the axial load capacity of a 203 × 203 UC 60 of 3.1 m height assuming that the loading acts at an effective eccentricity of 100 mm in the y–y direction at both ends (assume $p_y = 275$ N/mm²).

Solution
From section tables, $A = 75.8$ cm², $r_y = 5.19$ cm, $Z_y = 199.0$ cm³
$\therefore L/r_y = 3100/51.9$ and, noting from *Table 25* that strut curve c is appropriate corresponding value of p_c from *Table 27c* = 200 N/mm²
$\therefore P_c = 200 \times 75.8 \times 10^{-1} = 1516$ kN
$M_{cy} = 1.2 \times 275 \times 199.0 \times 10^{-3} = 65.7$ kN m
and from eqn (6.11)

$$\frac{F}{1516} + \frac{F \times 0.1}{65.7} + \frac{1}{2}\frac{F}{1516} \times \frac{F \times 0.1}{65.7} = 1$$

gives $F = \underline{418 \text{ kN}}$

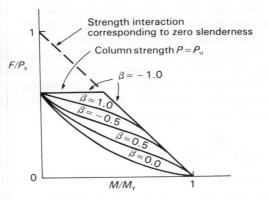

Fig. 6.9 Effect of moment gradient β on elastic limit interaction.

Example 6.4
What is the capacity of the column of example 6.3 for buckling about the major axis, assuming that the loading acts at an effective eccentricity of 100 mm from the column faces such as to induce double curvature bending?

Solution
From section tables, $r_x = 9.98$ cm, $S_x = 652.0$ cm^3
$L/r_x = 3100/89.8 = 34.52$ and noting from *Table 25* that strut curve b is appropriate from *Table 27b*, $p_c = 256$ N/mm^2
$\therefore P_c = 256 \times 75.8 \times 10^{-1} = 1940.4$ kN
From *Cl. 4.2.5*, $M_{cx} = 275 \times 652.0 = 179.3$ kN m
Total effective eccentricity $= D/2 + 100$ mm $= 204.8$ mm
and since $\beta = -1.0$ (double curvature) from eqn (6.12), $m = 0.43$, giving $\overline{M}_x = 0.43 \times F \times 0.205$ kN m
$\therefore$ from eqn (6.11)

$$\frac{F}{1940.4} + \frac{F \times 0.205 \times 0.43}{179.3} + \frac{1}{2} \frac{F}{1940.4} \times \frac{F \times 0.205 \times 0.43}{179.3} = 1$$

gives $F = \underline{894\ \text{kN}}$

This second example shows the benefit of using the m-factor to allow for the shape of the moment diagram since using $m = 1$ gives $F = 510$ kN, i.e. recognition of the less severe effect on overall buckling strength leads to almost a 40% gain in design capacity.

6.2.2 Case (2): Lateral–torsional buckling

The type of behaviour described above will normally occur only for I- and H-sections bent about their minor axis, for torsionally stiff sections such as tubes, or for strong-axis bending of I- and H-sections when the possibility of out-of-plane deformation is eliminated by the presence of an effective system of lateral bracing. I- or H-sections bent about their major axis will normally collapse by buckling in a mode that involves a combination of weak-axis bending and twisting; such behaviour is directly analagous to the lateral–torsional buckling of beams discussed in Chapter 5.

The elastic lateral–torsional buckling of beam-columns may be analysed in a manner very similar to the approach adopted for beams [2]. For sections having the normal proportions of columns, manipulation and simplification of the analysis results in an expression for the combination of axial load F and major-axis moment M (assumed for the present to be uniform along the member's length) that is analogous to eqn (6.10).

$$\frac{F}{P_{cry}} + \frac{M}{M_E (1 - F/P_{crx})} = 1.0 \tag{6.13}$$

However, in eqn (6.13) P_{cry} is now the critical load for buckling as a strut about the minor-axis and M_E is the critical moment for lateral–torsional instability under pure moment. Thus eqn (6.13) correctly represents the two extreme cases corresponding to $M = 0$ and $F = 0$. The amplification factor in the second term allows for the enhancement of the applied end moments in the manner shown in Fig. 6.6; the importance of this effect depends upon the member's major-axis slenderness, i.e. it is dependent upon P_{crx}.

When the problem is considered as one of the true ultimate strength of the member, analysis and test data show that eqn (6.13) provides a reasonable fit to the results providing P_{cry} and M_E are replaced by the strut and beam strengths determined from section 4.2.2 and eqn (5.1) respectively. Since the value of F will be limited to P_{cy}, which must be much smaller than P_{crx}, the effect of the amplification factor may be neglected, leading to the design condition of *Cl. 4.8.3.3.2*.

$$\frac{F}{P_{cy}} + \frac{M}{M_b} = 1.0 \tag{6.14}$$

Once again the strength of members subjected to unequal end moments M βM is rather higher. This effect may be approximated closely by replacing M in eqn (6.14) with an equivalent value $\overline{M} = mM$ with the value of m being obtained from eqn (6.12). Since a reduced moment is being used to check overall stability, it will also be necessary to ensure against local overstressing at the more heavily loaded end using eqn (6.4) or eqn (6.3).

Example 6.5
What is the capacity of the column of the previous example for buckling about the minor axis?

Solution
In this case it is first necessary to determine the member's lateral–torsional buckling strength as a beam M_b using the procedures of section 5.
From *Cl. 4.3.7.5*, $\lambda_{LT} = nuv\lambda$
which, using $u = 0.9$, $x = 14.1$, $n = 1.0$ according to *Cl. 4.3.7.6*,
λ = 59.7 from Example 6.3 and $v = 0.852$ from *Table 14*, gives
$\lambda_{LT} = 1.0 \times 0.9 \times 0.852 \times 59.7 = 45.8$
From *Table 11* corresponding value of $p_b = 248$ N/mm²
$M_b = 248 \times 652 \times 10^{-3} = 162$ kN m
From Example 6.3, $P_{cy} = 1516$ kN

$$\therefore \text{ using eqn (6.14)}, \frac{F}{1516} + \frac{F \times 0.20 \times 0.43}{162} = 1$$

giving $F = $ __840 kN__ (cf. 894 kN for major axis buckling)

In this example the ratio $P_{cy}/P_{cx} = 0.8$ and so both checks are necessary. Only two factors contributed to the different values of P_c: the value of λ and the change in the column curve. Other factors which could affect this include end restraint and intermediate bracing that is effective in one plane only, since both of these would lead to the use of different effective lengths in the two planes.

6.2.3 Case (3): Biaxial bending

The most general type of beam–column problem, which automatically incorporates the two previous cases, is the biaxially loaded member of Fig. 6.5 (iii). Even in the elastic range, analysis of the problem is extremely complex and explicit closed-form solutions cannot be obtained [8, 9]. Thus design equations must be based on an intuitive extension of the procedures of the two previous sections, properly checked against numerical and experimental data [2]. It is therefore convenient to discuss the basis for the design approach of BS 5950 in this case from a more qualitative standpoint.

The main features of the design of a beam–column may conveniently be displayed on a three-dimensional interaction diagram of the type shown in Fig. 6.10. In this, each of the three axes corresponds to one of the load components: compression F, major-axis moment M_x or minor-axis moment M_y. A safe design is one which may be represented by a point inside the appropriate failure surface. Because the exact form of the interaction varies with the slenderness of the member, the shape of this surface will be a function of a member's slenderness, with very stocky members being associated with a convex interaction of the type already illustrated in Fig. 6.1 for $L/r \to 0$. When one load component is absent the 3-D surface becomes a 2-D plane, for example when only F and M_x are present a safe design is one that plots below the curve joining the end points on the F and M_x axes appropriate to the member's slenderness.

For the full biaxial problem of Fig. 6.5 (iii), *Cl. 4.8.3.3.2* gives the design condition as

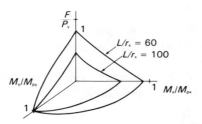

Fig. 6.10 Interaction surface for slender beam columns.

$$\frac{\overline{M}_x}{M_{ax}} + \frac{\overline{M}_y}{M_{ay}} \not> 1 \tag{6.15}$$

in which $\overline{M}_x$ = equivalent uniform moment about the x–x axis
$\overline{M}_y$ = equivalent uniform moment about the y–y axis
M_{ax} = buckling moment for combined axial load F and major-axis moment M_x
M_{ay} = buckling moment for combined axial load F and minor-axis moment M_y.

Equation (6.15) therefore locates a point in the M_x, M_y plane of Fig. 6.10, the end points of the curve defining this point having previously been determined by separate consideration of the F, M_x and F, M_y interactions. For simplicity the M_x, M_y interaction is taken as linear although some evidence exists to suggest that this is actually convex. In determining the quantities M_{ax} and M_{ay} the procedures reflect the different possible modes of failure illustrated in Fig. 6.5 (i) and 6.5 (ii) and described in sections 6.2.1 and 6.2.2.

Thus M_{ax} must be taken as the lower of

$$M_{ax} = \left[\frac{1 - F/P_{cx}}{1 + \frac{1}{2} F/P_{cx}}\right] M_{cx} \tag{6.16a}$$

or $M_{ax} = M_b(1 - F/P_{cy})$ \hfill (6.16b)

in which P_{cx} = axial strength for x–x buckling,
M_{cx} = in-plane bending strength about the x-axis,
P_{cy} = axial strength for y–y buckling,
M_b = lateral-torsional buckling resistance moment.

The first of these governs failure in the plane of the applied moments (Fig. 6.5 (i)), while the second controls out-of-plane buckling (Fig. 6.5 (ii)). In the most general case both must be checked since it will not be known in advance which will govern. However, when $P_{cx} > 1.5 P_{cy}$ the second condition will normally govern. Assuming equal degrees of end fixity in both planes (so that $\ell_x \simeq \ell_y$), the value of P_{cx} for 'normal sections' will often be found to exceed comfortably that of P_{cy}, with the result that eqn (6.16(b)) will more often control.

For moments about the minor axis, M_{ay} must be taken as

$$M_{ay} = \left[\frac{1 - F/P_{cy}}{1 + \frac{1}{2} F/P_{cy}}\right] M_{cy} \tag{6.17}$$

since only in-plane failure (Fig. 6.5 (i)) is possible, because as the member is being bent about its weak axis there is no tendency for it to fail by buckling in the plane at right angles.

An alternative, simpler but more conservative expression, which is analogous to eqn (6.3), is also permitted. This is

$$\frac{F}{Ap_c} + \frac{\overline{M}_x}{M_b} + \frac{\overline{M}_y}{p_y Z_y} \not> 1 \tag{6.18}$$

in which p_c = compression strength (lower of the values for x-x and y-y buckling)

A = gross cross-sectional area

Because eqns (6.15) and (6.18) use equivalent moments $\overline{M}$, a separate check against exceeding the local capacity of the member at its most heavily loaded cross-section is also necessary. This is achieved by using either eqn (6.4) or eqn (6.3) in the form of a pure strength check, i.e. M_{cx} should always be the pure cross-sectional bending capacity.

Example 6.6
Check the ability of a 3.1 m long 203 × 203 × UC 60 of grade 43 steel to carry a compressive load of 340 kN, assuming that this acts at effective eccentricities of 100 mm from the column face such as to produce single curvature bending about the yy-axis and double curvature bending about the xx-axis.

Solution
P_{cy} From section tables, $A = 75.8$ cm^2, $r_y = 5.19$ cm
Take $p_y = 175$ N/mm^2
$\lambda = 3100/519 = 59.7$
Corresponding p_c from *Table 27c* = 200 N/mm^2
$\therefore P_{cy} = 200 \times 7580 \times 10^{-3} = \underline{1516 \text{ kN}}$

P_{cx} From section tables, $r_x = 8.98$ cm
$\lambda = 3100/8.98 = 34.52$
Corresponding p_c from *Table 27b* = 256 N/mm^2
$\therefore P_{cx} = 256 \times 7850 \times 10^{-3} = \underline{1940.4 \text{ kN}}$

M_{cy} From section tables, $S_y = 302.8$ cm^3
$M_{cy} = 275 \times 302\,800 \times 10^{-6} = \underline{83.27 \text{ kN m}}$

M_{cx} From section tables, $S_x = 652.0$ cm^3
$M_{cx} = 175 \times 652\,000 \times 10^{-6} = \underline{179.3 \text{ kN m}}$

M_b Using *Cl. 4.3.7.5*, $u = 0.9$, $x = 14.1$
$\lambda/x = 59.7/14.1 = 4.23$
From *Table 14*, corresponding $v = 0.852$
$\lambda_{LT} = 1.0 \times 0.9 \times 0.852 \times 59.7 = 45.8$
From *Table 11*, corresponding $p_b = 248$ N/mm^2.
$M_b = 248 \times 652\,000 \times 10^{-6} = \underline{162 \text{ kN m}}$

M_{ax} From eqn (6.16a), $M_{ax} = 179.3 \times \dfrac{1 - 340/1940.4}{1 + \frac{1}{2} \times 340/1940.4}$
$\qquad = 136$ kN m
From eqn (6.16b), $M_{ax} = 162\,(1 - 340/1516) = 126$ kN m
$\therefore M_{ax} = \underline{126 \text{ kN m}}$ and minor-axis resistance controls (both checked because $P_{cy} > 2/3\,P_{cx}$)

M_{ay} From eqn (6.17), $M_{ay} = 83.27 \times \dfrac{1 - 340/1516}{1 + \frac{1}{2} \times 340/1516}$
$\qquad = \underline{58.2 \text{ kN m}}$

$\overline{M}_x$ Since $\beta = -1$ from eqn (6.12), $m = 0.43$
$\overline{M}_x = 340\,(100 + \frac{1}{2} \times 209.6) \times 0.43$
$\qquad = \underline{29.9 \text{ kN m}}$

$\overline{M}_y$ $\overline{M}_y = 340 \times 100 = 34.0$ kN m $(m = 1.0)$

Using eqn (6.15), $\dfrac{29.9}{126} + \dfrac{34.0}{58.2} = 0.237 + 0.584$

$$= \underline{0.821} \qquad \text{Satisfactory}$$

Check local strength at most heavily loaded cross-section; this will be at either end where the loads are $F = 340$ kN, $M_x = 69.6$ kN m and $M_y = 34.0$ kN m

Using eqn (6.5), $n = 340/(275 \times 75.8 \times 10^{-1}) = 0.163$

$S_{rx} = (1 - 2.5 \times 0.163^2) \times 652 = 609$ cm^3

Check $1.2 Z_x = 1.2 \times 581.1 = 697$ cm^3 $\qquad$ Satisfactory

$S_{ry} = (1 - 0.5 \times 0.163^2) \times 302.8 = 299$ cm^3

Check $1.2 Z_y = 1.2 \times 199.0 = 238.8$ cm^3 $\qquad$ Controls

$M_{rx} = 275 \times 608 \times 10^{-3} = 167$ kN m

$M_{ry} = 275 \times 238.8 \times 10^{-3} = 65.7$ kN m

Take $z_1 = 2.0$ and $z_2 = 1.0$

$$\left(\frac{69.6}{167} \right)^2 + \left(\frac{34.0}{65.7} \right) = 0.174 + 0.517$$

$$= 0.692 \qquad \text{Satisfactory}$$

Section is satisfactory for both overall buckling and local capacity. Alternatively, using eqns (6.18) and (6.3)

$$\frac{340}{75.8 \times 200 \times 10^{-1}} + \frac{29.9}{162} + \frac{34.0}{275 \times 199 \times 10^{-1}} = 0.224 + 0.184 + 0.621$$

$$= 1.029 \text{ and section is unsafe for overall buckling}$$

$$\frac{340}{75.8 \times 275 + 10^{-1}} + \frac{69.6}{179.3} + \frac{34.0}{65.7} = 0.163 + 0.388 + 0.518$$

$$= 1.069 \text{ and section is unsafe for local strength}$$

This biaxial example does, of course, incorporate all of the component problems covered in the earlier examples. In practice, where the requirement is usually one of checking the adequacy of a trial section rather than one of determining the precise load-carrying capacity, use of the simpler eqns (6.18 and 6.3) will generally prove easier. However, in the trial section just fails to meet these requirements (as is the case in this example), then recourse to the more exact provisions of eqns (6.15 and 6.4) may well enable that section to be used.

REFERENCES

1. Horne, M.R. *Plastic Theory of Structures* 2nd edn. Pergamon Press, Oxford (1979).
2. Chen, W.F. and Atsuta, T. *Theory of Beam-Columns*, vols. 1 and 2. McGraw-Hill, New York, (1976 and 1977).
3. BCSA/Constrado. *Structural Steelwork Handbook*. British Constructional Steelwork Association and Constrado, London (1978).
4. Morris, L.J. and Randall, A.L. *Plastic Design*. Constrado, London (1975).
5. Neal, B.G. *Plastic Methods of Structural Analysis*, 2nd edn. Chapman & Hall, London (1963).

6. Trahair, N.S. *The Behaviour and Design of Steel Structures*. Chapman & Hall, London (1977).

7. Johnston, B.G. *Guide to Stability Design Criteria for Metal Structures*, 3rd edn. John Wiley & Sons, New York (1976).

8. Culver, G.C. Exact solution of the biaxial bending equations. ASCE, *J. of Structural Division* **92** (ST2), 63–83 (April 1966).

9. Culver, G.C. Initial imperfections in biaxial bending. ASCE, *J. of Structural Division* **92** (ST3), 119–35 (June 1966).

EXAMPLES FOR PRACTICE

1. Use eqn (6.3) to determine the major axis moment that can safely be carried by a 254 × 254 UC 89 in Grade 43 steel that is already subjected to a tension of 1450 kN.

[166 kN m]

2. Compare the answer to question 1 with the result obtained using eqn (6.4) in conjunction with the formulae of the *Structural Steelwork Handbook*.

[209 kN m]

3. A 203 × 203 UC 52 is subject to an axial tension of 1125 kN. Assuming Grade 43 steel, can it also withstand moments of 53 kN m and 14 kN m about its major and minor axis respectively?

[Yes, assuming eqn (6.4)]

4. Determine the compressive load that can be carried by a 406 × 178 UB 60 in Grade 43 steel over a height of 5.6 m, assuming that it is braced against out-of-plane failure and that the maximum moment about its major axis is 72 kN m.

[1380 kN]

5. Determine the load-carrying capacity of a 305 × 305 UC 118 of effective height 3.6 m in Grade 43 steel, assuming it to be an external column in a simply connected frame with beam reactions of 105 kN.

[3200 kN]

6. Determine the end moments that can safely be carried by a 305 × 127 UB 37 in Grade 43 steel, assuming that these are in the ratio 1 : 0.3 and that they produce bending about the section's major axis. Take the member length as 4.8 m and allow for the presence of a 205 kN compressive load.

[10 kN m]

7. Check the ability of a 203 × 203 UC 60 of height in Grade 43 steel to carry the following load combination:

F (compressive) 750 kN
M_x 52 kN m   (top)   0.0 kN m   (bottom)
M_y 13.8 kN m   (top)   11.0 kN m   (bottom)

[Satisfactory]

7
Joints

Previous chapters have dealt with the design of different types of member such as beams and columns, with little consideration of the ways in which these are attached to one another to form a structure. However, many fabricators would argue that the economics of a steel structure are much more dependent upon the types of joint used than upon the sizes of the members. The basis for this lies in the fact that typical material costs represent only about 25–50% of the overall cost for the steelwork. It is thus not uncommon for the fabricator's own design staff to suggest modifications to joint details; providing the integrity of the structure is retained this is normally acceptable, since they are better placed to appreciate the equipment available and the effects on cost of various alternatives. Indeed, in some cases joint design is left entirely to the fabricator with the designer of the main structure supplying details of the loads which each connection must transmit together with any particular requirements, for example to provide adequate lateral restraint to the end of a beam.

Connections may involve the use of bolts (of which there are several different types) or welds, or a combination of both; rivets, although they are found in older structures such as railway bridges, are rarely used nowadays. Either type may be used for connections made in the fabricating shop; site connections will usually be bolted. Although it is possible to weld on site, the process is expensive since it requires special staging to provide a working platform, protection from the weather is necessary, the welds must be inspected and problems of access may arise since welding is much easier in certain positions, e.g. from above (downhand).

7.1 METHODS OF MAKING CONNECTIONS

7.1.1 Bolts

Three classes of bolt are in common use in the UK, although other types such as fitted bolts, are available to special order. These are:

(1) ISO metric black hexagon head bolts to BS 4190.
(2) ISO metric precision hexagon head bolts to BS 3692.
(3) High strength friction grip bolts to BS 4395.

For convenience these are usually referred to as 'black', 'precision' and 'HSFG' respectively.

Black bolts (which may be either black or bright in appearance) are made to less-stringent tolerances than are precision bolts; both are normally used in clearance holes 2 mm larger in diameter than the nominal bolt size. Although the former are available in three strength grades and the latter in ten, Grade 4.6 black bolts and Grade 8.8 precision bolts are the norm. (The first number is one tenth of the minimum UTS in kg/mm², while the product of both numbers gives the minimum yield stress in kg/mm². For example, Grade 4.6 min. UTS, $10 \times 4 = 40$ kg/m² min. yield stress $= 4 \times 6 = 24$ kg/mm².)

HSFG bolts are made from high-tensile steel and are tightened sufficiently with special torque spanners to produce a predetermined shank tension, thereby enabling additional shear resistance to develop between the connection plates as a result of friction. Installation is therefore a more critical operation and BS 4604 covers this as well as providing details of suitable design procedures. Guidance on the design of connections using HSFG bolts is available in a CIRIA Technical Note [1].

Although bolts are manufactured in a large range of diameters and lengths, certain sizes are 'preferred' and are therefore more readily available. Table 7.1 lists these for the case of black bolts. Preferred sizes for grade 8.8 and HSFG bolts are similar except that for the latter, M22, M27 and M33 are also included. Readers requiring further information on bolts are referred to part 1 of the BCSA publication on fasteners [2].

7.1.2 Welds

A weld is produced by passing a high current, typically between 50 and 400 amperes, through an electrode or filler wire so as to produce an arc which completes the path from the power source through the specimen to earth. Sufficient heat is produced – temperatures reached in the arc range between 5000°F and 30 000°F (2800–16 700°C) – to melt both the electrode and the parent metal so that the plates being welded fuse together on cooling. Typical specimens cut from welds are shown in Fig. 7.1. Possible embrittlement of the welded area is avoided by ensuring that while hot it is surrounded by an inert gas. This is provided by means of a substance called flux, either directly from the electrode as a core or coating, or, when bare wire is being used, in powder form.

Although welded joints produce cleaner lines, thereby avoiding possible corrosion traps, they generally require tighter tolerances than equivalent bolted joints. Also, the reduced preparation and handling must be set against the costs of the skilled labour required for the fabrication and subsequent inspection. Because of the obvious difficulty in checking the adequacy of a

Table 7.1
Manufacturers' recommended range of black hexagon head bolts to BS 4190 Grade 46, standard x and short o thread lengths, after reference [2].

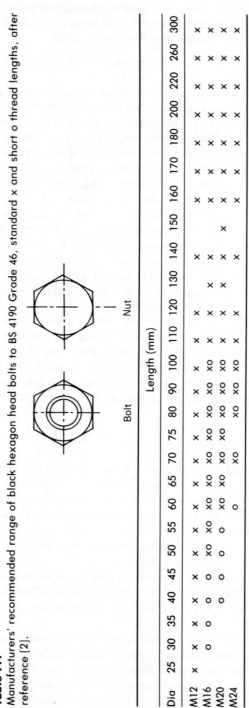

Dia	25	30	35	40	45	50	55	60	65	70	75	80	90	100	110	120	130	140	150	160	170	180	200	220	260	300
M12	x	x	x	x	x	x	x	x	x	x	x	x	x	x	x	x		x		x	x	x				
M16		o	o	o	o	xo	xo	xo	xo	xo	xo	xo	xo	xo	x	x	x	x		x	x	x	x	x	x	x
M20			o	o	o	xo	o	xo	xo	xo	xo	xo	xo	xo	x	x	x	x		x	x	x	x	x	x	x
M24							o	o	o	o	xo	xo	xo	xo	x	x		x	x	x	x	x	x	x	x	x

Length (mm)

Bolt   Nut

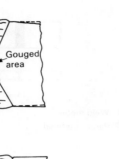

(a)

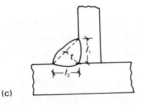

(b)

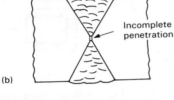

(c)

Fig. 7.1 Cross-sections of the main types of structural welds. t = throat thickness; ℓ_1 = vertical leg length; ℓ_2 = horizontal leg length. (*After ref. 3.*)

weld simply by visual means, inspection using more sophisticated methods, including x-ray, magnetic particle inspection (MPI), and ultrasonics, is normally employed. Full details of these techniques are provided in the appropriate British Standards; these are listed by Pratt [3]. In certain cases destructive tests on sample welds may be necessary as specified in BS 709.

Several different welding processes are available for the fabrication of structural steelwork. Probably the most widely used is the manual metal arc process (MMA); others include various automatic and semi-automatic processes such as CO_2, submerged arc and, where large deposition is required, electroslag. Full descriptions of these, together with guidance on the selection of the best process for a particular application, are available in section 7 of [3].

Figure 7.2 illustrates the two types of weld in common use for structural steelwork. For butt welds the weld metal is placed between the edges of the plates, whereas for fillet welds the weld metal is located on the faces of the plates. Various details, namely arrangements of the welds and corresponding edge preparations of the plates, are possible, especially when large welds are required. BS 5135 provides details of these as well as listing the agreed

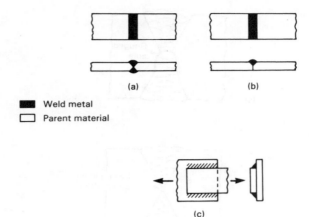

Fig. 7.2 Basic types of weld: (a) full penetration; (b) partial penetration; (c) fillet weld. (*Kulak, Adams and Gilmor, 1985.*)

symbols used to specify them on drawings. In certain cases where automatic or semi-automatic processes are to be used some modification may be permitted, providing all parties are agreeable; such agreement is usually based on procedural trials.

Butt welds may be either 'full penetration' or 'partial penetration' as shown in Fig. 7.2. The latter type are useful where access from both sides is impractical, although this does, of course, result in some eccentricity in the weld area which should be properly allowed for in design. Full penetration butt welds are designed on the basis of equivalence to the parent plate using the design strength of the parent metal, whereas partial penetration butt welds are assumed to possess an area corresponding to the depth of penetration only as explained in *Cl. 6.6.6*. Although full-penetration butt welds are structurally the most efficient (because they enable the full strength of the original cross-section to be utilized), the amount of fabrication involved even for the most usual type of double-V edge preparation tends to make them expensive. They should therefore be used only when circumstances really warrant it. For partial-penetration, single-V butt welds the efficiency as defined by the ratio of the axial stress in the plates to the maximum stress in the weld (allowing for bending effects) varies between about 20 and 60%, as plate thickness increases from 10 mm to 40 mm.

The load-carrying capacity of a fillet weld is obtained as the product of the throat area and the design strength of the weld p_w as given in *Table 36*. For symmetrically disposed fillet welds of the type shown in Fig. 7.2(c), *Cl. 6.6.5.1* permits p_w to be taken as the design strength of the parent material providing certain conditions on electrode type, throat thickness and stress conditions are observed. Strictly speaking p_w should be a function of the direction of loading [4], with transversely loaded welds being stronger than comparable longitudinal welds. However, since this increased strength

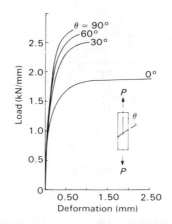

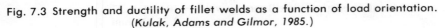

Fig. 7.3 Strength and ductility of fillet welds as a function of load orientation.
(Kulak, Adams and Gilmor, 1985.)

is obtained at the expense of ductility as shown in Fig. 7.3, BS 5950 follows
most other codes in specifying a single value. Figure 7.4 shows how for a 90°
fillet weld the effective throat size is determined as the dimension 'a' subject
to an upper limit of 70% of the effective leg length.

The values of p_w provided are based on experimental data [5] and
correspond to $0.47 \times$ UTS of the weld metal. They take into account the
simplifications inherent in basing the design of fillet welds on the average
stresses in the weld throat. Weld groups subject to a complex stress system
should be designed using a 'vector sum' approach as indicated by *Cl. 6.6.5.5*
such that the resultant stress does not exceed p_w. Useful comments on the
implementation of this approach are available in [6].

Example 7.1
Two plates are connected by means of a pair of fillet welds as shown in Fig. 7.5.
Assuming Gr. 43 material and electrodes to *Cl. 6.6.5.1* of BS 5950, what size
welds are required in order that a tensile force equal to the full strength of plate B
can be developed?

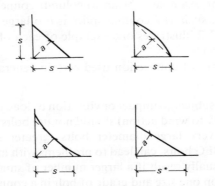

Fig. 7.4 Definition of throat sizes for 90° fillet welds.

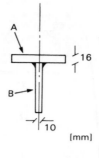

Fig. 7.5

Solution
From *Table 6*, for 16 mm material p_y = 275 N/mm²
∴ tensile strength of plate B per unit width = 10 × 1 × 275
                                       = 2.75 kN/mm
From *Cl. 6.6.5.1*, weld strength   = $2ap_w$
which, taking p_w as 215 N/mm² from *Table 32* = 2a × 215
    ∴ 2a × 215 = ~~2400~~ 2750
                a = 6 mm
and effective throat size of each weld = 6 mm.

Thus, providing the sum of the throat thicknesses of a pair of symmetrically disposed fillet welds slightly exceeds the thickness of the connected plate, the connection will permit full tensile load transfer. *Clause 6.6.5.1* actually permits p_w to be taken as p_y providing the two are equal; application of this rule to the present example would thus give the slightly lower value for *a* of 5 mm as being satisfactory.

7.2 GENERAL PRINCIPLES OF CONNECTION DESIGN

Structural connections are required when two different members must be joined together, for example a beam-to-column connection, or when an individual member, such as a column splice is too large for complete shop fabrication. Table 7.2 illustrates one example of each of the main types of steelwork connection.

Whatever the form of connection used certain general design principles should be observed:

(1) Connections subject to impact or vibration or load reversal (other than that due solely to wind action) should not use bolts in clearance holes.
(2) The use of very large diameter bolts (greater than about M30), especially HSFG bolts, can lead to problems with installation; a better design will usually result if a larger number of smaller bolts are used.
(3) Standardize on one size and grade of bolt in a connection and limit as far as possible the number of different sizes and grades in the structure.

Table 7.2
Examples of the main forms of steelwork connection (*Bates, Constrado Publications.*)

Type	Use
1	Beam to beam
2	Beam to column (transmits shear only)
3	Beam to column (full moment connection)
4	Truss connection

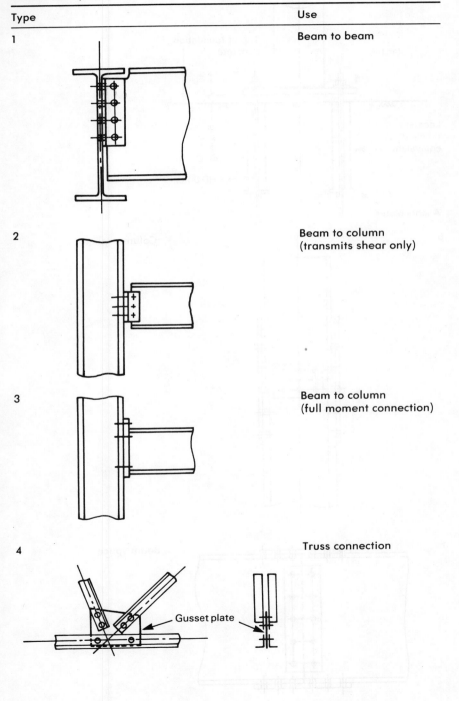

Gusset plate

Table 7.2 – *continued*

Type	Use
5	Column baseplate

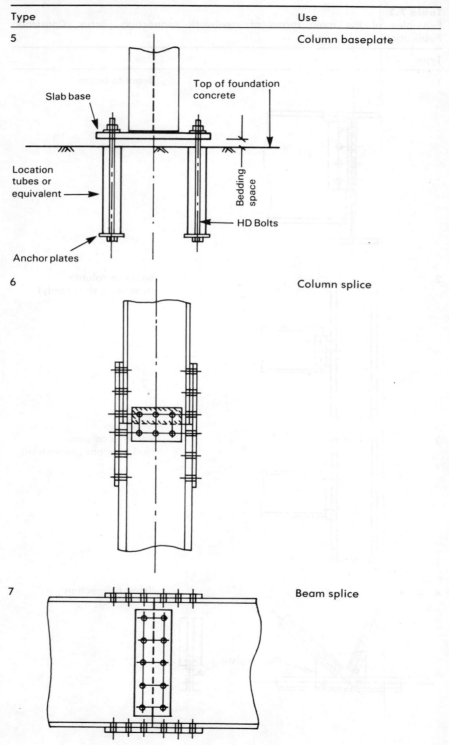

Slab base

Top of foundation concrete

Location tubes or equivalent

Bedding space

HD Bolts

Anchor plates

| 6 | Column splice |

| 7 | Beam splice |

Where different grades are required the inadvertant use of a lower grade bolt in place of the specified higher grade may be avoided by adopting different sizes, for example M16 black bolts and M20 grade 8.8 bolts.

(4) Before specifying HSFG bolts check for possible problems with installation and inspection.

(5) For welded joints subject to fatigue, as in crane rails, check Part 10 of BS 5400; try to avoid the lower class detail, i.e. those with poor fatigue performance.

(6) Do not specify larger fillet welds than are necessary. Avoid butt welds, which require expensive preparation, if fillet welds of a reasonable size would suffice.

(7) Consider the number of workshop operations required, for example for a member of all-welded construction apart from one end connection that requires drilling the cost of the separate process will be excessive.

(8) Avoid connection plates which require a large number of cuts. For trusses the gusset plates may be omitted altogether in certain circumstances and the joints made directly to the member.

(9) Avoid unnecessary splices in columns; unless the potential material savings are large or special factors are present; it will often be cheaper to run the heavier section through.

7.3 MODES OF FAILURE FOR FASTENERS

7.3.1 Bolts

Inspection of the example connections of Table 7.2 shows that the actual loading on the bolts will be either shear, tension or a combination of the two. For most forms of simple connection it is customary to design the bolts for shear only. The basic connection problem is therefore as shown in Fig. 7.6 in which several bolts in line are each subjected to a shearing action at the plate interface. Except in the case of long joints, defined by *Cl. 6.3.4* as exceeding 500 mm in length, the load on each bolt may be assumed equal.

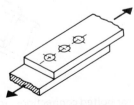

Fig. 7.6 Bolts in shear. (*Kulak, Adams and Gilmor, 1985.*)

This leads to the four possible types of failure shown in Fig. 7.7 only one of which actually depends upon bolt strength. The first of these – tearing at the net section of either plate – has already been covered in Chapter 3. Shearing of the plate beyond the end fastener should not occur providing the end distance exceeds $1.25d$ or $1.40d$ as defined by *Cl. 6.2.3*. This leaves the two most important failure modes: shearing of the bolt itself or bearing of the plate immediately behind the bolt, as illustrated in Figs. 7.7(c) and 7.7(d) respectively.

In determining the shear capacity of a bolt it is important to distinguish between the two cases:

(1) at least one shear plane passes through the threaded portion;
(2) threads do not occur in the shear plane.

Although case (1) is much more common, higher strengths can be developed for case (2) and this is recognized by permitting the use of the shank area A in

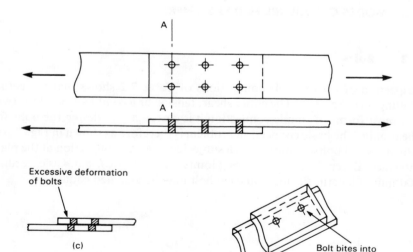

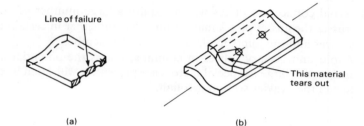

Fig. 7.7 Basic failure modes for bolted connections. (a) Tension at net section AA; (b) end failure of plate; (c) shear of bolts; (d) bearing.

such cases. For case (1) the area available to resist shear A_s will be the tensile area A_t. Thus the shear capacity P_s of one bolt in a condition of single shear as illustrated in Fig. 7.7 is given by *Cl. 6.3.2* as

$$P_s = p_s A_s \tag{7.1}$$

in which p_s = shear strength obtained from *Table 32*.
and A_s = shear area obtained from *Cl. 6.3.1*

The values given for p_s are the lesser of 0.69 times the yield strength or 0.48 times the ultimate strength of the fastener.

Bolts passing through more than two plates will possess a higher shear capacity since the total shear will be divided between the interfaces. As an example, the bolts in the web cover plates of the beam splice shown in Table 7.2 will be in double shear and the appropriate shear area for use in eqn (7.1) will therefore be $2A_s$. Caution is necessary, however, if the bolts pass through a total thickness of material significantly in excess of the bolt diameter as bending of the bolt will reduce the available shear capacity as explained in *Cl. 6.3.5*. Since this reduction does not become effective at thicknesses of less than $5d$ it will not often be required.

Bearing failure occurs when the bolt bites into the rear edge of its hole causing elongation and eventual tearing. Unless a low-strength bolt is used with higher-strength plates then the governing factor will be the bearing strength of the weakest connected ply, given by *Cl. 6.3.3.3*

$$P_{bs} = dt p_{bs} \tag{7.2}$$

in which d = effective, i.e. nominal diameter of the bolt
t = thickness of connected ply
p_{bs} = bearing strength of the connected parts obtained from *Table 33*.

Equation (7.2) assumes that sufficient material is present between the back face of the hole and the end of the plate. If this is less than twice the bolt diameter then bearing capacity must be reduced *pro rata*.

The values given for p_{bs} are based on considerations of serviceability, since actual bearing failure of the plate occurs at such high stresses that deformations will have become unacceptably large at a much earlier stage. For bolts in clearance holes the figure of 0.65 (ultimate strength + yield strength) used in *Table 33* reflects the approximate dependence of a suitable figure on the mean of the ultimate tensile stress and the yield stress.

For any given situation the strength of a bolt will clearly be the lesser of its capacities in shear and bearing. For the usual arrangements of plate thickness relative to bolt diameter the following inequality holds for Grade 4.6 bolts:

strength in single shear < strength in double shear < strength in bearing

The reason that bearing will not normally be critical is the extremely high values of p_{bs} given in *Table 33*. For Grade 8.8 bolts, for which much higher shear strengths are specified, the relative positions of bearing and double shear will often be reversed.

In the same way that fasteners should not be placed too near the ends of the connected plate they must also be suitably spaced both from each other and from the edges of the plates. The rules given in *Cl. 6.2* are based on several practical considerations. These include the provision of sufficient space between bolts to permit proper tightening, limiting the distance between bolts in compressive regions both to avoid buckling, and to avoid corrosion by ensuring adequate bridging of the paint film between plates [7].

Example 7.2
Calculate the strength of the bolts in the lap splice shown in Fig. 7.8 assuming the use of M20 grade 4.6 bolts in 22 mm clearance holes and Grade 43 plate.

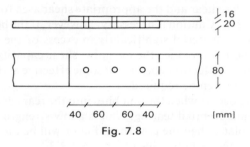

Fig. 7.8

Solution
Bolts are in single shear, from eqn (7.1) shear capacity per bolt
$\quad$ = 160 × 245 N = 29.2 kN
Bearing capacity of thinner plate per bolt, from eqn (7.2)
$\quad$ = 435 × 20 × 16 N = 139 kN
The full value is appropriate since the end distance $e \not< 2d$.
Clearly capacity is controlled by strength in shear. Therefore joint capacity in tension as governed by bolt strength = 3 × 39.2 = 118 kN

Moment resisting beam-to-column connections often contain regions in which the bolts will be required to transfer load by direct tension, such as the upper bolts in the end plate connection shown in Table 7.2. The capacity P_t of such bolts is determined from the equivalent of eqn (7.1) with tensile area A_t as specified in BS 3643 and tensile strength as given in *Table 32*. One rather contentious issue in the design of such connections concerns the additional forces induced in the bolts as a result of so-called 'prying action'. If one of the connected plates is sufficiently flexible that it may deform appreciably as illustrated in Fig. 7.9, then some allowance for the resulting bending of the bolts would appear to be in order. One suggestion [6] is that the nominal bolt forces be scaled up by the factor

$$\left(\frac{3b}{8a} - \frac{t^3}{20} \right) \tag{7.3}$$

However, *Cl. 6.3.6.2* treats prying in an alternative simpler fashion in that the tensile strengths provided in *Table 32* anticipate actual bolt forces being

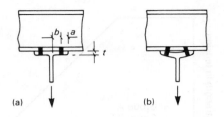

Fig. 7.9 Prying action causing bending of tension bolts passing through a flexible flange. (a) Rigid flange; (b) flexible flange. (*Kulak, Adams and Gilmor, 1985.*)

somewhat greater than those being designed for, i.e. conservative values are specified. Because of the importance of prying action in certain forms of connection, for example end plate beam-to-column joints as shown in Table 7.2, designers should try to make sensible allowances for it when selecting their connection design model [8, 9].

Where both shear and tension are present in the bolts, as with the upper bolts in the bracket connection of Table 7.2, then their combined effect may conveniently be assessed from a suitable interaction diagram. *Cl. 6.3.6.3* of BS 5950 specifies a trilinear diagram of the type shown in Fig. 7.10; this is represented by

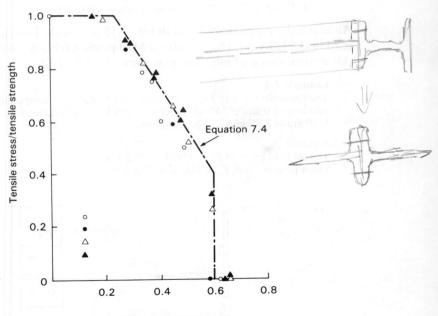

Fig. 7.10 Trilinear interaction curve for bolts under combined tension and shear, comparison with test data. (*From ref. 8.*)

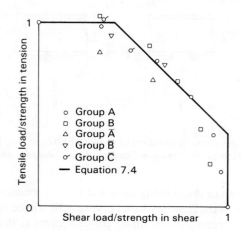

Fig. 7.11 Comparison of test results of ref. 10 for M20 black bolts in shear and tension with eqn (7.4), evaluation based on experimentally obtained values of P_s and P_t.

$$\frac{F_s}{P_s} + \frac{F_t}{P_t} \not> 1.4 \qquad\qquad (7.4)$$

in which F_s and F_t are the applied shear and tension
and P_s and P_t are the shear and tension capacities

Although the experimental data in Fig. 7.10 are for the higher grades of bolt used in the USA [8], recent tests on M20 grade 4.6 black bolts [10] support this general shape of interaction as shown by Fig. 7.11.

Example 7.3
The tee-stub shown in Fig. 7.12 is part of a beam-to-column connection which is required to transfer 350 kN in tension and 110 kN in shear. Check whether four M20 grade 8.8 bolts will be adequate.

Solution
Tensile load per bolt F_t = 350/4 = 87.5 kN
Shear load per bolt F_s = 110/4 = 27.5 kN

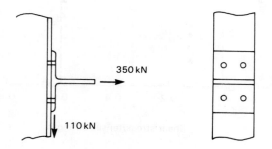

Fig. 7.12

From eqn (7.1), using $A_s = A_t$ (assumes shear plane passes through threads)

$P_s = 375 \times 245$ N $= 91.9$ kN
$P_t = 450 \times 245$ N $= 110.3$ kN

Check eqn (7.4)

27.5/91.9 + 87.5/110.3 = 0.30 + 0.79
                                    = 1.09   Satisfactory

7.3.2 Bolted connections using HSFG bolts

Figure 7.13 illustrates the type of load–deflection curve obtained from a typical test on an HSFG bolted connection loaded in shear. Of particular importance is the plateau corresponding to the load level at which slip between plies occurs, since this is absent for normal shear/bearing type connections. In BS 5950, ordinary parallel shank friction grip fasteners (except when used in long joints) are designed on the serviceability condition of slip presented as an ultimate check. Since such connections will have slipped into bearing at some stage between working and ultimate load a bearing capacity check is also necessary. By limiting the slip coefficient μ to a maximum of 0.55 adequate shear capacity at failure is ensured automatically (except in certain instances of long joints for which the check of *Cl. 6.4.2.3* is also necessary). For waisted-shank fasteners BS 5950 regards slip as 'failure'; since such connections must be designed on a non-slip basis the bearing check is unnecessary.

The slip resistance of parallel shank fasteners P_s, is given by *Cl. 6.4.2.1* as

$$P_s = 1.1 K_s \mu P_0 \tag{7.5}$$

in which P_0 = maximum shank tension from BS 4604
         μ = slip factor $\not> 0.55$
         K_s = 1.0 for fasteners in clearance holes (lower values are necessary in the case of oversize or slotted holes)

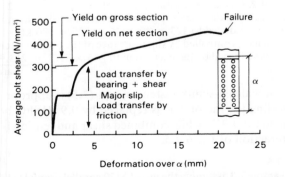

Fig. 7.13 Overall behaviour of a friction type connection showing effect of slip. (*Kulak, Adams and Gilmor, 1985.*)

Table 7.3
Recommended slip factors according to BS 5400: Part 3

Surface condition	Recommended slip factor*
Weathered steel clear of all mill scale and loose rust	0.45
Grit or shot blasted and with any loose rust removed	0.50
Blast cleaned and sprayed with aluminium	0.50
Blast cleaned and sprayed with zinc	0.40
Treated with zinc silicate paint	0.35
Treated with etch primer	0.25

* All slip factors should be reduced by 10% when higher grade bolts to BS 4395: Part 2 are used.

Slip factors should normally be determined from friction tests of the type specified in BS 4604. However, for general-grade fasteners and untreated surfaces the traditional μ-value of 0.45 is permitted. Table 7.3, which lists values for different types of surface as currently recommended by the Bridge Code BS 5400, shows something of the range of slip factors obtainable in practice.

Bearing is checked using eqn (7.2) with p_{bs} replaced by p_{bg}, values being obtained from *Table 34*. These are higher than for joints using other types of bolts due to relaxation of any requirements on acceptable deformation, i.e. the check is on strength only since the bolts will actually be in bearing only once the serviceability limit has been passed. For end distances of less than $3d$ a *pro rata* reduction is required.

Example 7.4
Repeat Example 7.2 assuming the use of M20 parallel-shank HSFG bolts in clearance holes.

Solution
Take $\mu = 0.45$ and obtain $P_0 = 144$ kN from BS 4604, note that the joint has only one pair of surfaces in contact.
From *Cl. 6.4.2.1*, slip resistance per bolt $= 1.1 \times 1.0 \times 0.45 \times 144 = 71.3$ kN
From *Cl. 6.4.2.2*, bearing resistance per bolt $= 20 \times 16 \times 825$ N $= 264$ kN
This assumes an end distance $e \nleqslant 3d$. However, since $e = 40$ mm, bearing resistance should be reduced to $1/3 \times 40 \times 16 \times 825 = 176$ kN.
Clearly capacity is controlled by slip resistance of the bolts and tensile capacity of connection as governed by fastener strength $= 3 \times 71.3 = \underline{213.9 \text{ kN}}$
Satisfactory

In the case of waisted-shank fasteners, eqn (7.5) is appropriate for checking slip, providing the factor 1.1. is replaced by 0.9. HSFG bolts in tension may be designed for $0.9P_0$, while combined shear and tension is controlled by the linear interaction of *Cl. 6.4.5*.

Example 7.5
Repeat Example 7.3 assuming the use of M20 parallel shank HSFG bolts in clearance holes.
Taking $\mu = 0.45$ and $P_0 = 144$ kN from BS 4604, from *Cl. 6.4.2.1*

slip resistance per bolt = $1.1 \times 1.0 \times 0.45 \times 144 = 71.3$ kN
From *Cl. 6.4.4.2*, tensile capacity per bolt = $0.9 \times 144 = 129.6$ kN
Tensile load per bolt = $350/4 = 87.5$ kN
Shear load per bolt = $110/4 = 27.5$ kN
Check interaction equation of *Cl. 6.4.5*

$$\frac{27.5}{71.3} + 0.8 \; \frac{87.5}{129.6} = 0.39 + 0.54$$

$$= \underline{0.93} \qquad\qquad \text{Satisfactory}$$

Although lack of fit between the connected plates may affect the degree of preload that may be achieved in each bolt in a connection, experimental evidence [9] suggests that this will not necessarily impair the subsequent performance of that connection.

Tightening of HSFG bolts to produce a given preload is normally controlled by one of the following:

(1) Torque control – use a calibrated manual wrench or a power wrench set to cut-out at a given torque; the value of torque used must be related to the required preload.
(2) 'Turn of nut' – after preliminary tightening with an ordinary podger spanner sufficient to bring the surfaces into contact, the nut and bolt shank end are marked, the nut is then turned further relative to the shank – typically one half or three-quarters of a turn is used – to provide a tension which normally exceeds the minimum proof load of the bolt.
(3) Direct tension indication – a load-indicating washer such as Coronet, or load-indicating bolt (Lib) is used to provide a direct indication of bolt tension; the principle is one of tightening to provide the desired gap under the bolt head, i.e. either device squashes as tension is increased but in a controlled way.

Of the three methods the use of direct indication, although more expensive in that special washers or bolts are required, is now the most widely used [2] in this country, although turn-of-the-nut is popular in North America.

Tightening of HSFG bolts is discussed from a practical point of view in the paper by Burdekin [12].

Example 7.6
A 150×20 mm tie in Gr. 43 steel carrying 400 kN requires a splice within its length. Design a suitable arrangement using a single-sided cover plate and (a) bolts in shear, (b) HSFG bolts, (c) fillet welds.

Solution
(a) *Shear-type bolted connection*
Try M20 grade 4.6 bolts in 22 mm holes, use 2 rows of bolts.
From *Cl. 6.2.1*, minimum spacing = $2\frac{1}{2} \times 20 = 50$ mm
From *Cl. 6.2.2*, maximum spacing = $14 \times 20 = 280$ mm
From *Cl. 6.2.3*, minimum edge and end distance (assuming a sawn edge) = $1.25 \times 22 = 28$ mm
Try the arrangement shown in Fig. 7.14.

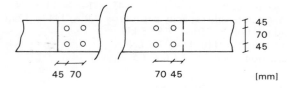

Fig. 7.14

From *Cl. 6.3.2*, capacity per bolt in single shear $= 160 \times 245$ N $= \underline{39.2 \text{ kN}}$

From *Cl. 6.3.3.3*, capacity per bolt in bearing in 20 mm plate $= 435 \times 20 \times 20$ N $= \underline{174 \text{ kN}}$

Therefore strength in shear governs and number of bolts required $= 400/39.2$ $= 10.2$

Use 12 bolts in 2 rows of 6.

Check total lap length against *Cl. 6.3.4*.

   Lap length $= 2 \times 45 + 5 \times 70 = 440$ mm

No reduction in bolt strength required as this is less than 500 mm.

Capacity of plate at net section, using $p_y = 265$ N/mm² from *Table 6* $= 265 \times (150 - 2 \times 22) \times 20$ N $= \underline{562 \text{ kN}}$ Satisfactory

An alternative would be to use grade 8.8 bolts for which

Capacity per bolt in single shear $= 325 \times 245$ N $= 91.9$ kN

Capacity per bolt in bearing in 20 mm plate is 160 kN as before.

Therefore strength in shear governs and number of bolts required $= 400/91.9$ $= 4.4$

Use 6 bolts in 2 rows of 3.

Since the single shear value governs the bolt capacity in both cases a more efficient joint would result if double-sided cover plates were used. These could be 10 mm thick, in which case six grade 4.6 bolts would suffice or four grade 8.8 bolts (bearing now controls).

(b) *Friction-type bolted connection*

Try M20 parallel-shank HSFG bolts in clearance holes, end and edge distances are basically as for shear type (a), except that *Cl. 6.4.2.2* requires a minimum end distance for fully effective bearing in the plate of $3 \times 27.5 = 82.5$ mm

From *Cl. 6.4.2.1* for 1 interface slip capacity per bolt $= 1.1 \times 1.0 \times 0.45 \times 144$ $= \underline{71.3 \text{ kN}}$

From *Cl. 6.4.2.2* bearing capacity per bolt $= 20 \times 20 \times 825$ N $= \underline{330 \text{ kN}}$

Therefore slip capacity governs and number of bolts required $= 400/71.3 = 5.6$

Use 6 bolts in 2 rows of 3.

Note A 45 mm end distance does not affect the capacity as bearing is not critical. Once again a more efficient arrangement would be to use a pair of cover plates to double the slip capacity in which case 4 bolts would be adequate.

(c) *Fillet welded connection*

In order to accommodate the welds on the flat surface of the tie it is necessary to use a cover plate of less than 150 mm width. Since its full cross-section will be effective a 100×20 mm plate should be adequate (this has approximately the same area as the net section area of the plate used in cases (a) and (b)).

From *Cl. 6.6.2.2*, minimum lap length $= 4 \times 20 = 80$ mm

From *Cl. 6.6.2.2*, if using longitudinal welds only $L \not< 100$ mm

From *Cl. 6.6.2*, end returns $\not< 2 \times$ leg length

Try 8 mm fillet welds

From *Cl. 6.6.5.3* throat thickness = $0.7 \times 8 = 5.6$ mm
Assuming the use of covered electrodes type E43, taking $p_w = 215$ N/mm² from *Table 36*.
Capacity of weld per mm run = $5.6 \times 215 = 1.20$ kN
Therefore required length = $400/1.12 = 333$ mm
Allowing for stop and start lengths according to *Cl. 6.6.5.2* gives a length of 333 $+ 2 \times 8 = 350$ mm
Therefore use 350 mm arranged as shown in Fig. 7.15.

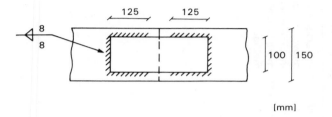

Fig. 7.15

7.4 EXAMPLES OF CONNECTION DESIGN

Actual connection design consists of identifying the load paths through the various parts of the connection, which must then be proportioned in such a way that an adequate margin against each possible type of failure (or limit state) is achieved. Usually this will require consideration of more than just the fastener-related modes described in section 7.3, since features such as the ability of gusset plates to withstand the forces induced by the members they connect, the need for column webs to resist high localized compression in beam-to-column connections, etc. must also be checked.

Because of the complexity of deciding on the exact pattern of loads and stresses within a joint, for example any attempt at rigorous analysis must include the effects of stress concentrations and localized plasticity, bolt slip, bolt preload, in-plane and bending action of the plates, local buckling, etc., it is usual to construct approximate models of joint behaviour [13–16]. Such 'models' seek to represent the main features in a manner that is sufficiently simple for rapid application in everyday design. Information of this type is not provided in BS 5950. However, for certain types of joint more than one acceptable model is available. This follows from the degree of simplification necessary to arrive at a workable design method being such that it can be arranged in a variety of ways, each of which fulfills the main structural requirements. Readers wishing to pursue this topic in greater depth should consult the appropriate specialist texts [8, 13–16].

Table 7.4
Beam-to-beam connections

Joint	Design basis	Comments	Ref.
7.4 (i)	Bolts 'A' carry vertical load in shear and bearing. Bolts 'B' carry same shear plus shear due to eccentricity e of bolts from face of cleat	Use vector sum method to allow for combined load-ing. Alternatively use a more 'exact' ultimate load method	[13] [17] [18]
7.4 (ii)	Bolts and welds carry vertical shear only		[13]

7.4 (iii)

As for 7.4 (ii)

Used where erection of (i) or (ii) may prove difficult

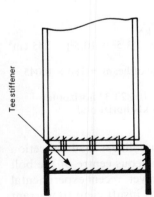

Tee stiffener

7.4 (iv)

Bolts 'A' carry vertical load, bolts 'B' transmit in shear the couple force due to the moment

Bolts 'C' could be omitted on the tension side (assuming no load reversal)

[13]
[14]

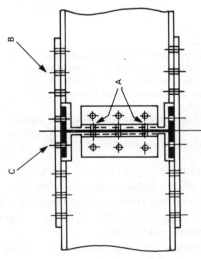

B

A

C

7.4.1   Beam-to-beam connections

Horizontal surfaces in structures, such as floors are often supported on a grid of intersecting beams. Such an arrangement necessitates connections of the type illustrated in Table 7.4. Types (i)–(iii) are suitable when only shear is being transferred, while the heavier type (iv) is an example of a moment-resisting beam-to-beam connection. The only aspect of the design of any of these connections which has not yet been explained is the effect of the eccentricity on the bolts B of type (i). Most authorities [8, 17] recommend the use of the 'vector sum' method (BS 5950 makes no specific recommendation); this is most easily appreciated by means of a worked example.

Example 7.6
Determine the force on the most heavily loaded bolt in the beam-to-beam connection illustrated in Fig. 7.16, assuming a beam end reaction of 180 kN.

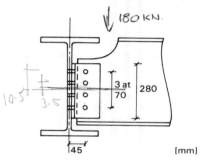

Fig. 7.16

Solution
Force per bolt due to vertical shear $= 180/4 = 45$ kN
I for bolt group about horizontal axis of bolt group $= 2[3.5^2 + 10.5^2] = 245$ cm^4
Z for outermost bolts $= 245/10.5 = 23.33$ cm^3
M due to eccentricity of line of bolts from centreline of beam $= 180 \times 0.045 = 8.1$ kN m
Force on outermost bolts $= 45$ kN vertical $+ 8.1 \times 10^2/23.33$ horizontal
$\qquad\qquad\qquad\qquad\qquad\quad = 45$ kN vertical $+ 34.76$ kN horizontal
Resultant force $= [45^2 + 34.76^2]^{\frac{1}{2}} = \underline{56.86 \text{ kN}}$

More sophisticated approaches, in which the true nonlinear load-deflection response of the bolts is used to locate the instantaneous centre of the bolt group by trial and error, are available [18, 19]. However, recent experimental work [20] suggests that the difference in accuracy is insufficient to warrant the additional calculations.

It is worth noting that when types (i) and (ii) involve similar depth beams then the ends of the secondary beam will have to be coped, i.e. part of one or both flanges will have to be removed as illustrated. Since this reduces the amount of lateral restraint provided [14], its effects upon the beam's overall bending strength as discussed in Chapter 5 must be taken into account.

Table 7.5
Beam-to-column connections suitable for 'Simple Construction'

Joint		Design basis	Comments	Ref.
7.5 (i)		Bolts 'A' carry the vertical load in shear and bearing. Bolts 'B' carry some shear plus shear due to eccentricity of bolts from face of cleat	Use vector sum method to allow for combined loading or use more accurate ultimate-load methods	[8] [13] [14] [17] [18]
7.5 (ii)		Bolts and welds carry vertical shear only		[8] [13] [14]

Table 7.5 – *continued*

Joint		Design basis	Comments	Ref.
7.5 (iii)	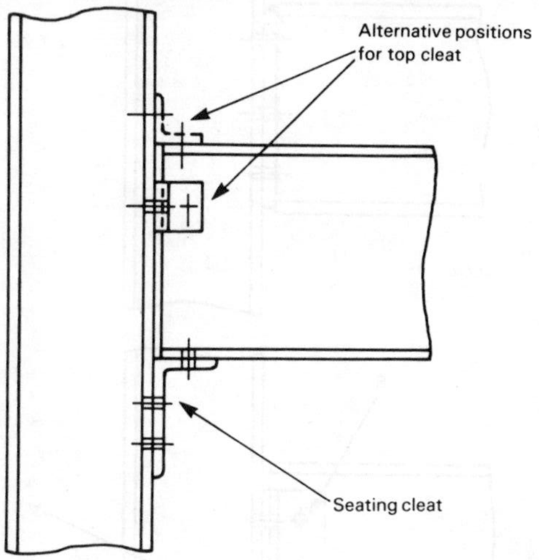	Seating cleat carries all vertical load. Top cleat provides lateral stability to beam. Design column bolts for vertical shear plus load due to eccentricity of centre of stiff bearing from column face	Shop bolted (or welded) seating cleat on column assists erection. Eccentricity may be ignored if small	[8] [13] [14]

Alternative positions for top cleat

Seating cleat

7.4.2 Beam-to-column connections – simple construction

Three examples of beam-to-column connection suitable for a frame design according to the principles of simple construction are shown in Table 7.5. Since their function is to transmit the beam reaction in shear into the column without developing significant moments, factors such as the provision of sufficient clearance between the column face and the lower flange should be properly considered. Seeking to give the end plate protection against possible damage in transit by extending it, perhaps accompanied by welding to the beam's bottom flange, results in significant changes in the way in which the joint behaves [14]. Types (i) and (ii) are the most commonly used, the choice between them depending upon the preferred method of shop fabrication, that is whether the beam should be provided with a bolted cleat or a welded end plate. Type (iii) possesses the advantage that the seating cleat may be used for 'landing' the beam during erection (for this it must, of course, be shop welded or bolted with the site joint being made to the beam).

Although the bolts on the column should be designed for the additional forces due to eccentricity e, this is often neglected when the effect is small. Use of this form of connection for very heavy loading may necessitate stiffening of the seating cleat. Both types (i) and (iii) provide some degree of tolerance for erection purposes via the clearance holes in the cleats, something that is lacking when type (ii) is used.

Example 7.7
Check the ability of the flush end plate beam-to-column connection illustrated in Fig. 7.17 to transfer a beam end reaction of 250 kN into the column. Both members are Grade 43 material, the end plate is $150 \times 280 \times 10$ mm, 6 mm fillet welds are used and the bolts are M20 grade 4.6.

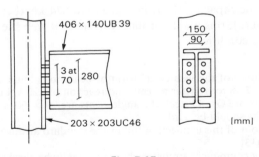

Fig. 7.17

Solution
The proportions of this connection follow the standard arrangement suggested in reference [13]. The following component strengths should normally be checked:

(1) Bolt group; (2) End plate; (3) Fillet welds; (4) Beam web.

(1) *Bolt group*
Bolt strengths according to *Table 32*:

Shear $= 160$ N/mm^2; bolt bearing $= 435$ N/mm^2
Bearing strength on plate according to *Table 33* $= 460$ N/mm^2
From *Cl. 6.3.2.2*, end distance for full bearing strength to be developed $= 2 \times 20 = 40$ mm
Since distance provided is 35 mm, bearing capacity of last row of bolts must be reduced *pro rata*.
By inspection bolt arrangement meets requirements of *Cl. 6.2* on spacing and edge distances.
From *Cl. 6.3.2*, taking A_s as the tensile area for threads in the shear plane
Capacity of bolts in single shear $= 8 \times 160 \times 245 = \underline{314 \text{ kN}}$

From *Cl. 6.3.3.2*, capacity of bolts in bearing on 10 mm end plate (column flange is 11.0 mm) $= 6 \times 435 \times 20 \times 10 + 2 \times 435 \times \frac{1}{2} \times 35 \times 10 = \underline{674 \text{ kN}}$
Therefore bolt group capacity is controlled by bolt shear.

(2) *End plate*
From *Cl. 4.2.3*, capacity of end plate in shear $= (0.9 \times 10 \times 280) \times 0.6 \times 275 \times 2 = \underline{832 \text{ kN}}$

(3) *Fillet welds*
From *Cl. 6.6.5.2*, effective length $= 2[280 - (2 \times 6)] = 536$ mm
From *Cl. 6.6.5.3*, throat thickness $= 0.7 \times 6 = 4.2$ mm
From *Table 36*, assuming E51 electrodes, capacity per mm run $= 4.2 \times 215$ N $= 0.90$ kN
$\therefore$ capacity of weld group $= 536 \times 0.90 = 482$ kN

(4) *Beam web*
From *Cl. 4.2.3* and taking $p_y = 275$ N/mm^2 from *Table 6*
Local shear capacity of beam web $= 0.6 \times 275 \times (0.9 \times 280 \times 6.3)$ N $= \underline{262 \text{ kN}}$
Summary of component capacities:

(1) Bolt group (shear)	314 kN	(3) Fillet welds	482 kN
(2) End plate	832 kN	(4) Beam web	262 kN

Therefore connection capacity is limited by the ability of the beam web to transmit shear; the connection is satisfactory for the 250 kN end reaction. Had the governing condition been the bolt group, a change to grade 8.8 bolts would improve the capacity of the bolt group to 674 kN (bearing in the end plate would govern); this would probably mean that web shear would again be the governing condition.

Example 7.8
Check the ability of the web cleats form of beam-to-column connection illustrated in Fig. 7.18 to transfer a beam end reaction of 120 kN into the column. Both members are Gr. 43 steel, the angle cleats are $90 \times 90 \times 8$ mm and M20 grade 4.6 bolts should be used.
The proportions of this connection follow the standard arrangements suggested in reference [13].
The following component strengths should normally be checked:

(1) Bolt group in beam web; (2) Bolt group in column flange; (3) Angle cleats in shear; (4) Angle cleats in bending.

(1) *Bolt group in beam web*
For 120 kN reaction, moment on these bolts $= 120 \times 0.05 = 6.0$ kN m
Using vector sum method to determine force on most heavily loaded bolt,
I for bolt group $= 2(3.5^2 + 10.5^2) = 245$ cm^4
Z for furthest bolts $= 245/10.5 = 23.3$ cm^3

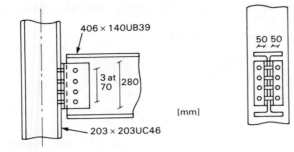

Fig. 7.18

Force on outermost bolt due to vertical shear = 120/4 = 30 kN
Horizontal force on outermost bolt due to moment = $6.0 \times 10^2/23.3$ = 25.8 kN
Resultant = $(30^2 + 25.8^2)^{\frac{1}{2}}$ = 39.6 kN
From *Cl. 6.3.3.3*, end distance for full bearing strength to be developed = 2 × 20 = 40 mm
Distance beyond hole in direction of resultant bolt force = 35 × 39.6/30.0 = 46.2 mm
By inspection, bolt arrangement meets requirements of *Cl. 6.2* on spacing and edge distances.
From *Cl. 6.3.2*, and taking A_s as the tensile area since shear plane passes through the threads
Capacity per bolt in double shear and = 2 × 160 × 245 N = $\underline{78.4\ kN}$
From *Cl. 6.3.3.2*, capacity per bolt in bearing in beam web = 20 × 6.3 × 435 N = $\underline{54.8\ kN}$
Since cleat thickness is 8 mm, bearing in this will be less critical.
Since capacity per bolt exceeds load on most heavily loaded bolt, group is satisfactory.
Capacity = (120/39.6) × 54.8 = $\underline{166.0\ kN}$

(2) *Bolt group in column flange*
From Cl. 6.3.2, capacity per bolt in single shear = 160 × 245 N = $\underline{39.2\ kN}$
From *Cl. 6.3.3.2*, capacity per bolt in bearing in 8 mm cleat = 20 × 8.0 × 435 N = $\underline{69.6\ kN}$
Actual capacity of last pair of bolts will be slightly less since end distance (vertical load) of 35 mm is less than the 40 mm required, ignore this as bearing is clearly not critical.
Therefore capacity of bolt group = 8 × 39.2 = $\underline{313.6\ kN}$

(3) *Angle cleats in shear*
From *Cl. 4.2.3*, shear capacity = 0.6 × 275 × (0.9 × 2 × 8 × 280) N
= $\underline{665.4\ kN}$

(4) *Angle cleats in bending*
Capacity of connection = 166.0 kN
Shear capacity of cleats = 665.4 kN
From *Cl. 4.2.5* since 166.0 < (0.6 × 580.7), take $M_c = p_y Z$
Gross I for cleats = $1.8 \times 28.0^3/12$ = 3293 cm⁴
less holes 2 × 1.8 × 2.2 × $(3.5^2 + 10.5^2)$ = $\underline{970\ cm^4}$
Net I for cleats = 2323 cm⁴
Z for cleat = 2323/14.0 = 165.9 cm³
∴ M_c = 275 × 165.9 N = 45.6 kN m

In terms of reaction at 50 mm eccentricity, this corresponds to a force of 45.6/0.05 = <u>912 kN</u>

Summary of component capacities:

(1) Bolt group in beam web	<u>166.0 kN</u>
(2) Bolt group in column flange	313.6 kN
(3) Angle cleats in shear	665.4 kN
(4) Angle cleats in bending	912.0 kN

7.4.3 Beam-to-column connections – continuous construction

This form of connection may be made in a great variety of ways, six of which are illustrated in Table 7.6. Before considering these in detail it will be useful to establish certain points relating to the design of moment-resisting connections in general:

(1) The beam end moments will also contribute to the shear force at the joint.

(2) Axial tension or compression may be present in the beam; its effect on connection design should be approached with caution since such forces may well be present only under certain conditions of loading.

(3) Tension in the beam will, as a result of rotation of the joint, produce additional moment. If this effect is significant, placing of the bolts symmetrically with respect to the resultant line of action of the applied forces enables them to be designed for tensile forces only, thereby assisting in keeping connection size reasonable.

(4) Compression in the beam, since it has the opposite effect, can lead to lighter connections.

(5) The compression zone of the column web should be checked for possible failure in local bearing and buckling (see Chapter 5); some stiffening may be necessary [8, 13, 14, 21, 23, 24].

(6) Moment connections to one column flange only require the attendant shear to be transmitted through the column web; again stiffening may be required [14, 21, 23].

Table 7.6 illustrates six examples of moment-resisting beam-to-column joints suitable for use in continuous construction. The most popular of these is type (i) the extended end-plate. Variants of this are possible in which the end-plate is made almost flush with the bottom of the beam (assuming downward loading on the beam) or even when it is effectively contained within the beam depth, although evidence [21]isuggests that for the latter case very thick plates are necessary to resist the induced moments (equal to the product of the beam flange force and its distance from the nearest row of bolts). A detailed treatment of end-plate connection design is provided in reference [25].

Example 7.9
Determine the capacity of the extended end plate beam-to-column connection illustrated in Fig. 7.19, assuming that both members are Grade 43 steel, the end

Table 7.6
Beam-to-column connections suitable for 'continuous construction'

Joint		Design basis	Comments	Ref.
7.6 (i)	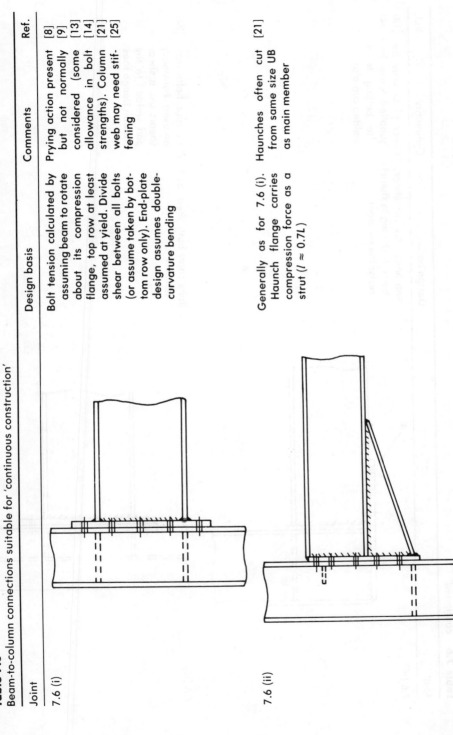	Bolt tension calculated by assuming beam to rotate about its compression flange, top row at least assumed at yield. Divide shear between all bolts (or assume taken by bottom row only). End-plate design assumes double-curvature bending	Prying action present but not normally considered (some allowance in bolt strengths). Column web may need stiffening	[8] [9] [13] [14] [21] [25]
7.6 (ii)		Generally as for 7.6 (i). Haunch flange carries compression force as a strut ($l \approx 0.7L$)	Haunches often cut from same size UB as main member	[21]

Table 7.6 – *continued*

Joint	Design basis	Comments	Ref.
7.6 (iii)	Web bolts take the shear, bolts 'A' (acting in shear) resist the moment	Cover plate may be supplied loose for site bolting to a welded cap plate	[18] [24]
7.6 (iv)	Bolts carry both shear and tension	Only four bolts may be used. For heavy shears use a shear pad welded to the column flange toes	[24]

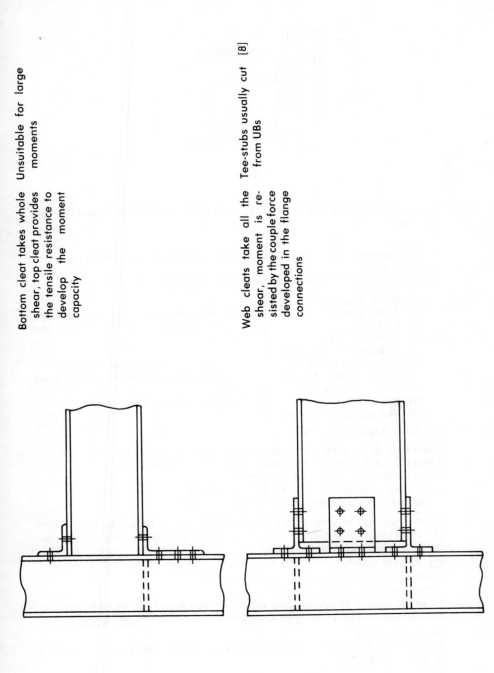

7.6 (v)

Bottom cleat takes whole shear, top cleat provides the tensile resistance to develop the moment capacity

Unsuitable for large moments

7.6 (vi)

Web cleats take all the shear, moment is resisted by the couple force developed in the flange connections

Tee-stubs usually cut from UBs [8]

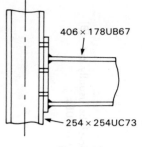

Fig. 7.19

plate is 200×20 mm, the beam flange welds are 10 mm fillet welds and the bolts are M20 grade 8.8.

Solution
The proportions of this connection follow the standard arrangement suggested in reference [13].
The following component strengths should normally be checked:

(1) Beam flange welds ⎫
(2) Tension bolts ⎪
(3) End plate ⎪ Compare with tensile force in beam flange
(4) Column flange ⎬ due to moment and axial load in the
(5) Column web in tension ⎭ beam.

(6) Column web in shear   Compare with shear in column web.
(7) Column web in buckling ⎱ Compare with compressive force in beam
(8) Column web in bearing ⎰ flange.
(9) Beam web welds ⎱ Compare with shear in beam.
(10) End plate bolts in shear ⎰

(1) Beam flange fillet welds

From *Cl. 6.6.5.2*, effective length $= 2\,[178 - (2 \times 10)] = 316$ mm
From *Cl. 6.6.5.3*, throat thickness $= 0.7 \times 10 = 7.0$ mm
From *Table 36*, assuming E51 electrodes,
Capacity per metre run $= 70 \times 215$ N $= 1.5$ kN
∴ capacity of weld group $= 1.5 \times 316 = \underline{474\ \text{kN}}$

Note Since sum of throat thickness (20 mm) exceeds beam flange thickness (14.3 mm) from *Cl. 6.6.5.1*, these welds are capable of transmitting a force equal to the tensile capacity of the flange.

(2) Tension bolts
From *Table 32*, tensile strength $= 450$ N/mm²
Tensile stress area $A_t = 245$ mm²
From *Cl. 6.3.6.1*, tensile capacity per bolt $= 450 \times 245 = 110.3$ kN
Tensile capacity of the four bolts in the group $= 4 \times 110.3 = \underline{441.2\ \text{kN}}$

(3) Beam end plate
From *Cl. 4.2.3*, capacity of end plate in shear $= (0.9 \times 200 \times 20) \times 0.6 \times 275 \times 2$ N $= \underline{1188\ \text{kN}}$

Since tensile capacity of bolts $= 441.2$ kN, this is less than 0.6×1188 and from *Cl. 4.2.5*, moment capacity of end plate $M_c = p_y S \not> 1.2 p_y Z$.
For a rectangular plate, $S = bd^2/4$ and $Z = bd^2/6$;
Therefore take $M_c = 1.2 p_y Z = 1.2 \times 275 \times (200 \times 20^2/6) = 4.40$ kN m

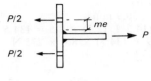

Fig. 7.20

Assuming double-curvature bending of the end plate [13], this moment capacity must be capable of resisting the couple due to the product of the force in the top row of bolts and the distance of the bolt centreline from the toe of the tension flange weld, *me* as indicated in Fig. 7.20.

$$\therefore \text{ capacity of end plate } = \frac{4.40 \times 2}{0.5 \times 0.045} = \underline{391.1 \text{ kN}}$$

(4) *Column flange*

A method of checking the ability of the column flange to withstand the tensile force produced by the four tension bolts is provided in reference [21]. This is used (in a modified permissible stress format) for several examples in reference [13]. Because of their length the calculations have not been included herein: interested readers should refer to reference [21] to confirm that the capacity of the column flange in bending = $\underline{611 \text{ kN}}$. Providing sensibly proportioned connections are used, such as a column flange thickness of approximately 70% of the end-plate thickness, this component is unlikely to prove critical.

(5) *Column web in tension*

Effective depth of column web in tension is obtained by taking a 60° dispersion from the two rows of tension bolts to the column flange root location (Fig. 7.21). Horizontal distance from centreline of bolts to column flange root

$$= \frac{100 - 8.6 - 2 \times 12.7}{2} = 33 \text{ mm}$$

Effective depth of web = $4 \times 33 \sqrt{3}$ = 228.6 mm
Effective area acting in tension = 228.6×8.6 = 19.7 cm^2
From *Cl. 4.6.1*, tensile capacity of column web = $19.7 \times 10^2 \times 275$ N
$$= \underline{541.8 \text{ kN}}$$

(6) *Column web in shear*

From *Cl. 4.2.3*, shear area = 8.6×254 = 21.8 cm^2
Shear capacity of web = $0.6 \times 275 \times 21.8 \times 10^2$ N
$$= \underline{360.4 \text{ kN}}$$

(7) *Column web in buckling*

From *Cl. 4.5.1.3*, stiff length of bearing, $b_1 = 10.9 + 2 \times 20$ = 50.9 mm
From *Cl. 4.5.2*, n_1 = 254 mm
From *Cl. 4.5.2*, $\lambda = 2.5 \times 200.2/8.6$ = 58.2
From *Table 27c*, p_c = 205 N/mm^2
Buckling capacity of web = $(50.9 + 254) \times 8.6 \times 205$ N
$$= \underline{537 \text{ kN}}$$

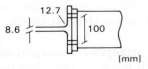

Fig. 7.21

(8) *Column web in bearing*
Stiff length of bearing = 50.9 mm
From *Cl. 4.5.3*, n_2 = 2(14.2 + 12.7) × 2.5 = 134.5 mm
Bearing capacity = (50.9 + 134.5) × 8.6 × 275 N
              = 438.5 kN

(9) *Beam web welds*
Effective length of 6 mm fillet welds = 2 × 360.5 = 721 mm
From *Cl. 6.6.5.1*, capacity per mm run = 0.7 × 6 × 215 N = 0.90 kN
Capacity of weld group = 0.90 × 721 = 649 kN

(10) *End plate bolts in shear*
Beam end shear is assumed to be resisted by the two bolts adjacent to the beam's
compression flange.
From *Cl. 6.3.2*, single shear capacity of one bolt = 375 × 245 N = 91.9 kN
From *Cl. 6.3.3.3*, capacity per bolt in bearing in 14.2 mm column flange = 20 ×
14.2 × 460 N = 130.6 kN
Capacity of the two shear bolts = 2 × 91.9 = 183.8 kN

Summary of component capacities:

(1) Beam flange welds	474 kN	(6) Column web in shear	360 kN
(2) Tension bolts	441 kN	(7) Column web in buckling	537 kN
(3) End plate	391 kN	(8) Column web in bearing	439 kN
(4) Column flange	611 kN	(9) Beam web welds	649 kN
(5) Column web in tension	542 kN	(10) End plate bolts in shear	184 kN

Items (3), (6) and (8) should exceed the demands made by the moment and
axial load transmitted by the beam. Although the case of shear in the column
web gives the lowest figure, for the case of double-sided connections, i.e.
beams on both column flanges, column shears will often be small. With
insignificant axial loads in the beam(s) the requirements for tensile and compressive capacity will, of course, be similar.

In situations where the moment at the joint exceeds the capacity of the
beam section, a haunched connection of the type shown in Table 7.6 as
7.6 (ii) may be used, a common example being the eaves of a portal frame (see
Chapter 8). Haunches may be made either from split UB sections or from
plate.

At a column cap the type of joint shown as 7.6 (iii) is suitable. An alternative, all-welded arrangement would be to run the beam through the connection and to use vertical stiffeners to extend the column flanges. A design
model based on North American practice is provided in reference [18] and
[24].

Although types (i) and (ii) are also suitable for beams framing into the
column web this may present difficulties if moment connections are required
on both axes. Type 7.6 (iv) represents one means of making such a joint by
employing tee-stiffeners to effectively move the connection to the column
face. Such stiffeners will, of course, also act to stiffen the column web against
major axis bending.

The top and bottom cleat arrangement used previously as a simple connection can be used to transmit moments providing the bottom cleat is made
much more substantial, which in turn will probably require stiffening of the

adjacent column web. A cleated connection capable of transmitting large moments is shown in 7.6 (vi). This uses tee-stubs cut from UBs as the flange connections. Since these are symmetrically loaded they deform less than the eccentrically loaded angles of type (v); they also permit the use of more bolts.

Not shown in Table 7.6 are any all-welded joints. Structurally, these represent the simplest form of moment-resisting beam-to-column connection. However, this must be balanced, not only against the need to employ site welding, but also against the generally rather higher degree of precision necessary in fabrication and fit-up. None the less, such connections are sometimes used in the UK; they are much more common in regions such as North America and Japan, where greater use is made of continuous construction. For a discussion of their design, which requires that careful consideration be given to factors such as ductility and the provision of adequate stiffening, the reader is referred to references [8, 14, 17, 18, 22–24].

7.4.4 Gusseted connections

In situations where a number of differently oriented members meet at a joint, such as in a truss or lattice girder, the transfer of forces between them may be achieved conveniently by means of a piece of plate termed a gusset. Figure 7.22 illustrates an example of the type often seen in fairly light roof trusses.

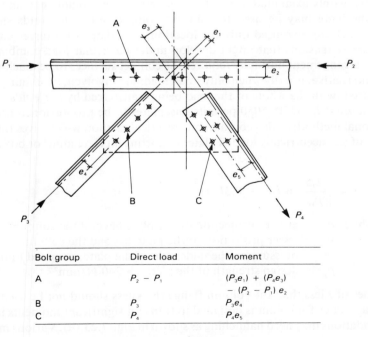

Bolt group	Direct load	Moment
A	$P_2 - P_1$	$(P_3e_1) + (P_4e_3)$ $- (P_2 - P_1) e_2$
B	P_3	P_3e_4
C	P_4	P_4e_5

Fig. 7.22 Gusseted connection

Ideally the centroids of all members, and thus the lines of action of the direct tensile or compressive forces in them, should intersect at a point. However, the practicality of actually making the connection may not permit this, in which case the effects of eccentricity of loading should be allowed for in its design. Although this is often neglected for light to medium trusses, proper allowance should be made when designing the connections between heavily loaded members, for example joints in large trusses fabricated from universal column sections of the sort used in power stations and bridges.

The gusset plate itself will normally be subjected to bending, shear and axial force components induced as a result of shear transfer through the fasteners from the members. Thus design of the actual gusset consists essentially of checking a number of critical sections, usually at bolt positions as indicated in Fig. 7.22. Even in cases where the load application is not symmetrical with respect to the gusset plate, as with a gusset connecting a series of single angles, the joint geometry is often such that out-of-plane bending is insignificant; it is therefore usual to neglect this in design.

7.4.5 Column baseplates

Transfer of column loads into masonry or concrete foundations usually requires the insertion of a steel plate between the two components if overstressing of the weaker foundation material is to be avoided. For columns carrying only axial load, direct bearing between the column end and the top of the plate may be used to transmit compression. The welds shown in Fig. 7.23 are then used only for location or perhaps to transfer any small shears or tensions that might develop under particular load combinations. This arrangement normally requires the contact surfaces to be machined. As an alternative, usually when only small loads are involved, machining may be omitted, with the whole of the load being transferred by the welds.

Clause 4.13 of BS 5950 permits baseplates to be proportioned using any rational method; it also contains an empirical method which gives the thickness of a concentrically loaded plate supporting an H, channel or box column as

$$t = \left[\frac{2.5}{p_{yp}} \, w \, (a^2 - 0.3b^2) \right]^{\frac{1}{2}} \tag{7.6}$$

in which a = greater projection of the plate beyond the column
b = lesser projection of the plate beyond the column
w = pressure on the underside of the plate, assumed uniform
p_{yp} = design strength of the plate, $\ngtr$ 240 N/mm²

Values of t less than the column flange thickness should not be used.

Baseplates for columns designed to transmit significant moments into the foundations may need haunching as shown in Fig. 7.23 (ii). Various methods [17] are available for assessing the pressures under the baseplate, the thick-

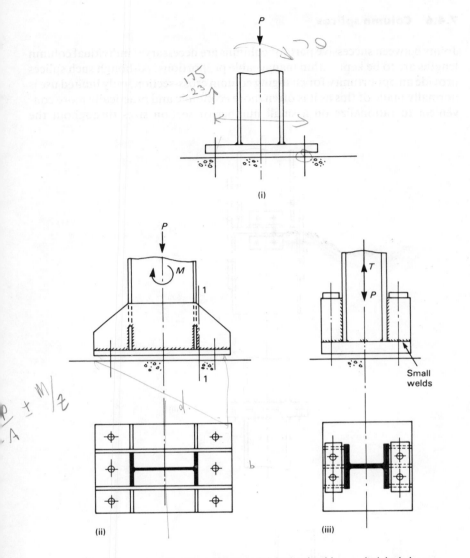

Fig. 7.23 Column bases: (i) slab base, (ii) haunched base, (iii) bolt boxes.

ness of which may be determined by treating it as a beam spanning between the haunch plates.

Cases where baseplates are required to transmit large tensile forces entail the use of very thick plates to resist the moments produced by the holding-down bolts. In combination with heavy welding this can lead to lamellar tearing [3] in the baseplate. One way of avoiding it is to modify the method of load transfer by using bolt boxes as shown in Fig. 7.23 (iii). Most of the load is now carried by the fillet welds between the boxes and the column flanges.

7.4.6   Column splices

Joints between successive parts of columns are necessary if individual column lengths are to be kept within manageable proportions. Although such splices provide an opportunity for changing column cross-section, only limited use is normally made of this as it is often more economic and practically more convenient to rationalize on a small number of section sizes throughout the

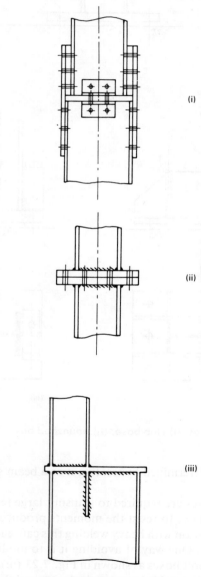

Fig. 7.24 Column splices.

project. Apart from special situations, for example where heavy additional loads must be carried over only a portion of the column's height as occurs with crane columns, it is usual practice to retain common outside dimensions over the full column height.

For cases of predominantly axial loading a nominal connection of the type illustrated in Fig. 7.24 (i) will normally prove adequate [13, 14]. Typical flange plate dimensions would be a projection on either side of the joint equal to the column flange width but not less than 300 mm, with a thickness equal to that of the upper column flange but not less than 12 mm. If different serial sizes are involved a machined division plate, its thickness being based on dispersion of load at 30°, is normal.

If load reversal is possible, end-plate connections provide a convenient solution as shown in Fig. 7.24 (ii). High-tensile or possibly HSFG bolts are usual and care is necessary in selecting material free from laminations for the end plate due to the tensile loading involved.

Connections between columns of very different size may be arranged as shown in Fig. 7.24 (iii); both faces of the division plate should be machined and its thickness will normally need to be at least 20 mm. The web stiffener, which assists in diffusing load into the lower column, should be of similar proportions to the upper column flange.

7.4.7 Beam splices

Long-span beams may require site connections between successive lengths. Figure 7.25 illustrates two basic forms of beam splice, both of which can have

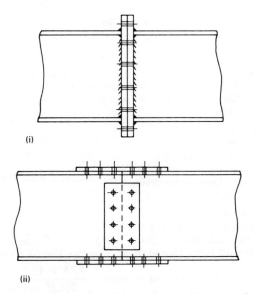

Fig. 7.25 Beam splices: (i) end plates, (ii) cover plate.

several variants. For the end-plate arrangement, design is similar to that discussed previously for the beam-to-column end plate, i.e. shear is assumed to be shared equally between all bolts with the moment being resisted by a group of tension bolts. The flange cover plates in Fig. 7.25 (ii) should be capable of transmitting the whole of the moment with the web bolts taking shear plus the secondary moment due to their eccentricity. For large beams, flange plates may be placed on both faces of the beam flanges; HSFG bolts will often be required if the number of bolts used is to remain reasonable. As an alternative, welded splices may be used to provide a particularly clean appearance. References [11] and [12] both provide detailed discussion of various design approaches together with example calculations for a number of different types of beam splice.

Example 7.10
Check whether the beam splice illustrated in Fig. 7.26 is capable of transmitting a moment of 150 kN m together with a shear of 250 kN m. Flange cover plates are 15 mm and web cover plates are 8 mm. All bolts are M20 general grade HSFG and all material is Grade 43.

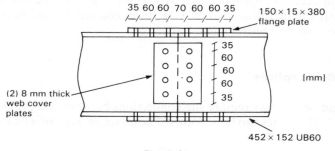

Fig. 7.26

Solution
The proportions of this connection are in accordance with those suggested by reference [13]. Design is based on the assumption that moment is transmitted entirely by the flange plates with the web plates carrying the whole of the shear. Items to be checked:

(1) Web splice bolts   ⎫
(2) Web cover plates in shear ⎬ Compare with resultant force due to
(3) Web cover plates in bending ⎭ shear + moment due to eccentricity

(4) Flange splice bolts   ⎫
(5) Flange cover plates in compression ⎬ Compare with flange force due to
(6) Flange cover plates in tension   ⎭ moment

(1) *Web splice bolts*
Vertical shear on bolt group = 250 kN
Moment due to eccentricity = 250 × 0.035 = 8.75 kN m
Force on outermost bolts: Due to shear   = 250/4 = 62.5 kN

$$\text{Due to moment} = \frac{8.75 \times 13.5 \times 10^2}{2(4.5^2 + 13.5^2)} \text{ N}$$

$$= 29.1 \text{ kN}$$

Resultant force $= (62.5^2 + 29.1^2)^{\frac{1}{2}} = 68.9$ kN
From *Cl. 6.4.2.1*, assuming $\mu = 0.45$ and noting that two interfaces are present
Slip resistance of one bolt $= 1.1 \times 1.0 \times 0.45 \times 144 \times 2 = 142.6$ kN
From *Cl. 6.4.2.2*, noting that $e = 35 \times (69.9/29.1) > 3d$
Bearing resistance in 8 mm beam web $= 20 \times 8 \times 825$ N $= 132$ kN
For the 16 mm of cover plate, $e = 35 \times (68.9/62.5) < 3d$, but capacity is still $>$ 132 kN

Therefore capacity of bolt group in shear $= \dfrac{250 \times 132}{68.9} = \underline{480.0 \text{ kN}}$

Since this exceeds 250 kN applied this item is satisfactory.

(2) *Web cover plates in shear*
From *Cl. 4.2.3* and *Table 6*, shear capacity of cover plates
$\quad = 0.6 \times 275 \times (0.9 \times 8 \times 340) \times 2$ N
$\quad = \underline{807.8 \text{ kN}}$ Satisfactory

(3) *Web cover plates in bending*
Since capacity of web splice bolts (480.0 kN) $< 0.6 \times$ shear capacity of web cover plates (485 kN),
from *Cl. 4.2.5*, take $M_c = p_y S \not> 1.2 p_y Z$
For a rectangular plate, $S = bd^2/4$ and $Z = bd^2/6$
$\therefore$ use $M_c = 1.2 p_y \, bd^2/6$ and shear capacity at eccentricity of 35 mm
$\quad M_c = 1.2 \times 275 \,[16 \times 340^3/12 - 16 \times 22 \times 2\,(45^2 \times 135^2)]/(170 \times 35)$
$\quad\quad = \underline{2115 \text{ kN}}$ Satisfactory

Note In determining Z, allowance has been made for the presence of holes.

(4) *Flange splice bolts*
Force taken by bolt group on either side of splice $= 150/0.455 = 330$ kN
From *Cl. 6.4.2.1*, taking $\mu = 0.45$ for one interface
$\quad P_s = 1.1 \times 1.0 \times 0.45 \times 144 = 71.3$ kN
From *Cl. 6.4.2.2*, for bearing in 13.3 mm flange plate, noting that $e = 35$ mm,
$\quad P_{bg} = 1/3 \times 35 \times 13.3 \times 825$ N $= 128$ kN
$\therefore$ capacity of bolt groups $= 6 \times 71.3$
$\quad\quad\quad\quad\quad\quad\quad\quad\quad\quad = \underline{427.8 \text{ kN}}$ Satisfactory

(5) *Flange plate in compression*
For top plate from *Cl. 4.7.4*, $P_c = A_g p_c \not> A_e p_y$
From *Cl. 3.3.1*, gross area $A_g = 22.5$ cm^2
but from *Cl. 3.3.3* for Gr. 43 material
$\quad A_e = 1.2 \times 15.9 = 19.08$ cm^2

Noting that close spacing of bolts will give a low slenderness so that $p_c \simeq p_y$,
Capacity of cover plates in compression $= 1908 \times 275 = \underline{525 \text{ kN}}$
$\quad\quad\quad\quad\quad\quad\quad\quad\quad\quad\quad\quad\quad\quad\quad\quad\quad\quad\quad\quad$ Satisfactory

(6) *Flange plate in tension*
From *Cl. 4.6.1*, $P_t = \underline{525.0 \text{ kN}}$ as before   Satisfactory

Summary of component capacities:

(1) Web splice bolts	480 kN
(2) Web cover plates in shear	808 kN
(3) Web cover plates in bending	2115 kN
(4) Flange splice bolts	428 kN
(5) Flange cover plates in compression	525 kN
(6) Flange cover plates in tension	525 kN

Items (i)–(iii) exceed the load produced by the shear while items (iv)–(vi) are capable of resisting the 330 kN flange force produced by the moment.

REFERENCES

1. Cheal, B.D. Design guidance for friction grip bolted connections. CIRIA Technical Note 98, London (1980).
2. Boston, R.M. and Pask, J.W. *Structural fasteners and their application*. BCSA, London (1978).
3. Pratt, J.L. *Introduction to the welding of structural steelwork*. Constrado, London (1979).
4. Butler, L.J. and Kulak, G.L. Strength of fillet welds as a function of direction of load. Welding Research Supplement, *Welding Journal* **36** (5) 321s–234s (May 1971).
5. Higgins, T.R. and Preece, F.R. Proposed working stresses for fillet welds in building construction. Welding Research Supplement, *Welding Journal*, 429s–32s (October 1968).
6. AISC. Design of fillet weld groups. *AISC Steel Construction* **13** (1) (1979).
7. Firkins, A. and Hogan, T.J. Bolting of steel structures. *AISC Steel Construction* **11** (3) (1977).
8. Fisher, J.W. and Struik, J.A.H. *Guide to design criteria for bolted and riveted joints*. John Wiley & Sons, New York (1974).
9. Morris, L.J. Connection design in the UK. In *ASCE Manual on Steel Beam-to-Column Building Connections* edited by Chen, W.F. In press.
10. Shakir-Khalil, H. and Ho, C.H. Black bolts under combined tension and shear. *The Structural Engineer* **57B**, 69–76 (December 1979).
11. Mann, A.P. and Morris, L.J. Lack of fit in steel structures. CIRIA Report No. 87, London (1981).
12. Burdekin, F.M. Tightening HSFG bolted joints. *Metal Construction*, 387–9 (July 1982).
13. Pask, J.W. *Manual on connections for beam and column construction*. BCSA, London (1982).
14. Hogan, T.J. and Thomas, I.R. *Design of structural connections* (standardized structural connections manual – Part B). Australian Institute of Steel Construction (1981).
15. Chen, W.F., ed. Steel beam-to-column building connections. In *Manual on Steel Beam-to-Column Building Connections*, ASCE, (in press).
16. Owens, G.W. *Design of structural steelwork connections*. Butterworths, London (1985).
17. McGuire, W. *Steel structures*. Prentice-Hall, Englewood Cliffs (1968).
18. Kulak, G.L. Adams, P.F. and Gilmor, M.I. *Limit states design in structural steel*. Canadian Insitute of Steel Construction, Ontario, (1985).
19. Crawford, S.F. and Kulak, G.L. Eccentrically loaded bolted connections. *Journal of the Structural Division, ASCE* **97** (ST3), 765–83 (March 1971).
20. Bahia, C.S. and Martin, L.H. Bolt groups subject to torsion and shear. *Proc. Inst. Civil Engrs* **69** (2), 473–90 (June 1980).
21. Horne, M.R. and Morris, L.J. *Plastic design of low-rise frames*. Granada, London (1981).
22. Chen, W.F. and Lui, E.M. Static web moment connections. In *ASCE Manual on Steel Beam-to-Column Building Connections* edited by Chen, W.F. In press.
23. Chen, W.F. and Lui, E.M. Flange moment connections. In *ASCE Manual on*

Steel Beam-to-Column Building Connections edited by Chen, W.F. In press.
24. American Society of Civil Engineers, Plastic design in steel. I *Manual 41* 2nd edn. ASCE (1971).
25. Cheal, B.D. Limit state design of bolted connections. CIRIA RP306, London (1984).

EXAMPLES FOR PRACTICE

1. What size fillet welds are required to attach a 150×12 mm flat bar hanger to the bottom flange of a 457×152 UB 74 so that the full tensile capacity of the hanger may be developed? Assume the use of Grade 43 steel and electrodes for which $p_w = 215$ N/mm².

   [7 mm throat size]

2. What is the capacity of an M16 Grade 4.6 bolt passing through a 12 mm plate and a 15 mm plate in (a) single shear, (b) bearing, (c) double shear assuming two 12 mm plates? Assume that the shear plane(s) pass through the threaded portion. State any conditions necessary for these strengths to be available.

   [25.1 kN, 50.2 kN, 86.4 kN end distance $\not< 32$ mm]

3. How many M24 Grade 8.8 bolts will be needed in a tension splice comprising two 16 mm cover plates on the longer leg of a $200 \times 100 \times 15$ mm angle in Grade 43 steel, if the full strength of the angle is to be developed?

   [6]

4. What is the shear load that can safely be carried by four M16 grade 8.8 bolts that are already carrying 30 kN each in tension? Assume that the bolts are in single shear.

   [136 kN]

5. What is the capacity of a group of four M20 parallel shank HSFG bolts in clearance holes assuming a slip coefficient between contact surfaces of 0.50?

   [327 kN]

6. Assuming that the bolt group in question 5 is subject to a shear load of 240 kN, what additional tensile load could it safely withstand?

   [156 kN]

7. How many M20 grade 4.6 bolts are needed in a tension splice on the longer leg of a $150 \times 90 \times 12$ mm angle if the member is carrying 60% of its axial capacity? Sketch a suitable arrangement.

   [10]

8. Repeat question 7 assuming the use of (i) M20 grade 8.8 bolts, (ii) M20 HSFG bolts (take $\mu = 0.45$). What length of 8 mm fillet weld would also be suitable?

   [4, 6, 285 mm]

9. A 610×305 UB 149 is to be connected to the flange of a 350×368 UC 153 by a pair of web cleats using six M20 grade 8.8 bolts in a single line on the beam web. Determine the resultant force on the most heavily loaded bolt if the vertical reaction on the beam is 540 kN. Assume 70 mm spacing between bolts and 50 mm eccentricity. Check whether this bolt is adequate.

   [109 kN, unsafe in bearing in the beam web]

10. Check whether an 8 mm thick end plate in grade 43 steel of less than the full beam

depth used in conjunction with ten M20 grade 4.6 bolts, would be a suitable connection between a 457 × 191 UB 74 beam and a 254 × 254 UC 89 column if the
beam end reaction is 300 kN.

[Suitable]

11. A pair of 90 × 90 × 10 mm angles are to be used as web cleats to form a connec-
    tion between a 610 × 229 UB 113 and a 254 × 254 × UC 132 using six M20 grade
    8.8 bolts in a single line in the beam web. Is this a safe arrangement for a beam
    reaction of 400 kN?

[Yes]

12. Check the suitability of a 16 mm thick extended end plate welded to a 406 × 178
    UB 54 with 6 mm fillet welds and fastened to the flange of a 254 × 254 UC 74
    with six M20 grade 8.8 bolts to carry a shear of 90 kN and a moment of 80 kN m.

[Suitable]

13. Check whether a flush end plate welded to both flanges of the beam but with all
    the bolts contained within the beam depth can be used as an alternative solution
    for question 12.

[Yes]

14. Design a baseplate for a 305 × 305 UC 198 assuming that the column has been
    designed as 'pin-ended'.

[700 × 700 × 50 mm with four M24 grade 4.6 nominal bolts
and 8 mm nominal fillet welds would be suitable]

15. A 457 × 152 UB 60 beam requires a splice at a point where the shear is 260 kN
    and the bending moment is 150 kN m. Assuming the use of 8 mm web cover
    plates and 15 mm flange cover plates with M20 general grade HSFG bolts, design
    a suitable joint.

[A possible arrangement would be 350 × 150 × 15 mm flange
plates, 340 × 140 × 8 mm web plates, 8 web bolts and a total of
24 flange bolts]

8
Continuous construction

The term 'continuous construction' was introduced in section 2.1 to describe the class of steel frames in which the types of joint employed were able to maintain virtually unchanged the original angles between adjacent members. Since such joints are capable of transmitting moments, the behaviour under load of these frames is more complex than that of the alternative 'simple construction'. In particular, member forces, i.e. moments, shears and thrusts, cannot now be obtained directly using simple statics. In principle, of course, utilizing continuity is structurally more efficient because of the greater participation of all parts of the structure in resisting the applied loads. Whether or not the resulting structure will actually be superior, that is more economic, more robust, etc., than the simply designed alternative is a complex question that was discussed briefly in section 2.2.

For simple design the procedures described in Chapters 3–6 enable individual members to be selected; they may also be used for members in continuous structures providing the internal forces in these members have been calculated properly. This may be done on an elastic basis using any suitable method, such as moment distribution, slope deflection, and matrix stiffness [1]ior, providing certain restrictions are observed, using plastic theory [2, 3]. The second approach utilizes the ductility of steel, as discussed in section 1.2, to permit redistribution of moments after the attainment of maximum capacity locally in the most highly stressed member. Since its use assumes the strength of the structure to be governed by a particular mode of collapse, it is necessary to eliminate the possibility of premature failure by other means, for example through the use of strict geometrical limits to control buckling.

Some indication of the types of joint suitable for use in continuous construction has already been provided in Chapter 7. Since these are normally more complex than those used for simple construction their use will involve more fabrication and they will therefore be more expensive. Erection on site may also be more difficult because of the tighter degree of fit-up required

Fig. 8.1 Typical portal frame structure during erection. (*Courtesy of Condor Midlands, Burton-on-Trent.*)

(more exact matching-together of the individual components). Thus continuous construction does have certain practical disadvantages which must be weighed against the more obvious structural advantages of using less steel and producing a generally stiffer, more robust structure.

Although special requirements may affect the choice for a particular structure, construction economics in the UK have tended to push the use of continuous construction into certain well-defined areas. Probably the most important of these is the use of portal frames of the type illustrated in Fig. 8.1 for low-rise industrial buildings such as factory units and warehouses. It has been suggested [4] that these consume something approaching one half of the UK civil engineering market for structural steelwork. As a result, their design has become a highly refined and competitive process. Other important areas include the use of continuous beams in situations where limiting deflections or minimizing construction depth are important – multistorey frames for which considerations of access and flexibility in utilizing the internal space make the use of bracing unacceptable, and for highway bridges where continuity over the supports provides a better riding surface.

8.1   ELASTIC DESIGN OF CONTINUOUS STRUCTURES

The design of continuous steel structures on an elastic basis consists essentially of applying the methods described in the preceding chapters using member forces calculated in a way that recognizes the effects of continuity. Because of the greater degree of interaction between different parts of the structure it may be more difficult to identify the exact load case corresponding to the most severe condition in each individual member. This is likely to be particularly true for members subject to combined loading for which design has to be based on some form of interaction approach, for example the support region of a continuous plate girder where coincident shear and moment values must be considered and beam-columns carrying compression and unequal end moments. Fortunately BS 5950 permits member forces to be obtained using linear elastic analysis, amplifying these where necessary to allow for instability effects. This has two important consequences for the designer: it enables him to sum the effects of different load cases using the principle of superposition and it gives him the opportunity to use standard frame analysis programs for extensive or irregularly shaped structures.

However, the Code gives little guidance on which arrangements of load are likely to be the most critical; *section 5.1.2* merely refers to vertical loads 'arranged in the most unfavourable but realistic pattern for each element'. Horizontal loads need only be considered to act in conjunction with full vertical load. An earlier draft [5] did attempt to be more specific, including suggestions in the Commentary for use when considering multistorey frames. Even the use of these does not guarantee that the most severe design condition for each member will be included [6]. When considering pattern loading it is not necessary to allow for variations in dead load as part of the pattern, i.e. γ_f should be taken as 1.0 for dead load throughout the structure.

Deflections under serviceability load conditions are generally subject to the same limits as for simply designed structures, as covered by *Cl. 2.5.1* and *Table 5*. An exception is made for portal frames and this is discussed in section 8.6.

Provided an elastically designed continuous structure contains only members of compact cross-section, see section 5.3.1, limited redistribution of moments is permitted. Thus within a beam for example, the elastic moment diagram may be modified by up to 10% of the peak elastic moment, providing, of course, the resulting moments and shears remain in equilibrium with the applied loads. This concept may be thought of as a very limited recognition of the potential that exists within continuous structures to withstand loads in excess of those that require full member bending strength only at the most critical location. Since this is possible only if unloading does not follow the attainment of this local maximum strength, some limitation on cross-sectional geometry is required; this is the reason for limiting the process to compact sections.

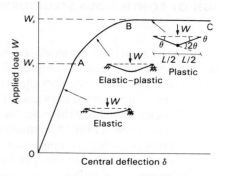

Fig. 8.2 Behaviour of a simply supported beam. (*BSC Teaching Project, Imperial College*, 1985.)

8.2 PLASTIC DESIGN OF CONTINUOUS STRUCTURES

8.2.1 Behaviour of continuous structures

The main differences between the behaviour of 'simple' and continuous steel structures can best be appreciated by considering a specific example.

Figure 8.2 shows the load versus central deflection relationship obtained from either a test or from a rigorous theoretical analysis for a simply supported beam. Three distinct phases may be identified:

(1) OA elastic – linear relation between load and deflection.
(2) AB elastic plastic – deflections increase at a progressively faster rate.
(3) BC plastic – growth of large deflections at sensibly constant load.

Detailed consideration of the distribution of bending strain and bending stress at the most severely loaded cross-section at mid-span reveals that during phase 2 yielding is developing, while phase 3 corresponds to a state of full local plasticity (strictly speaking this is true only if certain simplifications are used which make it rather easier to appreciate the basic features of inelastic bending, for example the real stress–strain curve is approximated as a bilinear elastic – perfectly plastic relationship). The attainment of a fully plastic cross-section in bending is termed the formation of a 'plastic hinge'. The reason for the first part of the name is self-evident; the assumption of perfectly plastic material behaviour beyond yield means that the effective modulus for the material will be zero and thus the cross-section's effective value of *EI* will be zero. The beam will now behave rather as if a real hinge had been introduced at mid-span in that it will become a mechanism unable to resist any further increase in load – hence the horizontal load–deflection curve.

The behaviour of a continuous structure, for example the two-span continuous beam of Fig. 8.3, will exhibit certain differences as indicated by the

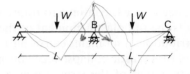

Fig. 8.3 Two-span continuous beam.

load versus mid-span deflection relationship shown as Fig. 8.4. In this case the beam's response up to the formation of a plastic hinge at the point of maximum moment, based on the elastic moment diagram, will be basically similar to that of the simply supported beam. However, because of the redundancy in the system the appearance of this hinge at the central support transforms the beam not into a mechanism but into a statically determinate structure. Although it will function somewhat differently, in that the additional moments developed as a result of the application of increased load will be distributed differently, i.e. the moment at the central support cannot increase, it will none the less be capable of withstanding extra load.

Thus continuous structures possess the ability to redistribute load from the most severely stressed locations (plastic hinges) to less highly stressed areas. Only after sufficient plastic hinges have formed to convert the original redundant structure into a progressively less redundant structure, then into a statically determinate structure and finally into a mechanism, does collapse occur. Of course, utilization of this ability requires that premature failure by other means, such as elastic or plastic instability, or material breakdown due to insufficient ductility, be prevented. Therefore use of 'plastic design', as the exploitation of this redistributive phase after the formation of the first plastic hinge is termed, involves making certain assumptions:

(1) The steel has adequate ductility as measured by the possession of a sufficiently long plastic plateau.

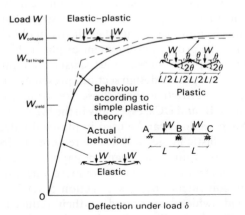

Fig. 8.4 Load–deflection curve for a statically indeterminate steel beam. (BSC Teaching Project, Imperial College, 1985.)

(2) Plastic hinges, once formed, continue to rotate at a sensibly constant moment (taken as the full plastic moment of the cross-section).

(3) Sufficient redistribution of moments can occur for the structure to fail by the formation of a plastic collapse mechanism.

(4) Instability effects, either of the structure as a whole, of individual members or of the component plate elements of a member, do not prejudice the formation of the collapse mechanism.

Much of the material in BS 5950 relating to plastic design is concerned with ensuring that these conditions are met. Readers who are unfamiliar with the methods of plastic analysis necessary for the determination of plastic collapse loads should consult a suitable textbook [2–4].

8.2.2 Requirements for the use of plastic design

The use of plastic design relies on the ability of steel to accept strains considerably in excess of yield, so that those regions of the structure in which plasticity develops (plastic hinges) can maintain their capacity to carry load. Thus ductile behaviour is required (a) from the steel so that yield may develop fully over member cross-sections and (b) from the members so that full redistribution of moments may occur. In order to achieve this, limits must be placed on the type of steel used and the proportions of the members employed. Other safeguards on ductile behaviour are that regions containing plastic hinges should be fabricated to a high standard as set out in *Cl. 5.3.7* of BS 5950 and that the structure be subject to predominantly static loading.

Plastic design is permissible for steel grades 43, 50 and 55 to BS 4360. If other grades are to be used they must satisfy the requirements on ductility given in *Cl. 5.3.3*. Possession of an adequate plastic plateau is clearly necessary for the development of yield over a cross-section. It is perhaps not immediately clear why an ultimate tensile strength significantly in excess of the yield strength is also required; this ensures that strain hardening will take place. Although this is not normally used explicitly in plastic design, its existence is essential for the proper development of plastic hinge action [2, 3].

Members containing plastic hinges must satisfy the limitations on flange and web proportions for plastic design sections presented in Table 8.1. These are sufficiently more restrictive than those for compact sections that a number of standard UBs and UCs, especially in the higher grades of steel, will not meet them. Such sections can be used in plastically designed structures only in areas where plastic hinges will not be required to form.

Use of plastic design requires that beams be capable of attaining and maintaining their full plastic moment capacity (reduced where necessary to allow for the effects of coincident shear, see section 5.3.1.2), and that beam-columns carry loads which correspond to their reduced plastic moment capacity, see section 6.1, etc. Since the required member strengths are prescribed in advance, it is necessary to ensure that premature failure by member

Table 8.1
Cross-sectional limits necessary to prevent local buckling in members required to participate in plastic hinge action (based on Table 7)

Type of element	Method of manufacture		Limiting proportions for plastic design	
		$p_y = 275$ N/mm²	$p_y = 355$ N/mm²	$p_y = 450$ N/mm²
Outstand element of compression flange	Welded b/T	≯ 7.5	≯ 6.5	≯ 6
	Rolled b/T	≯ 8.5	≯ 7.5	≯ 6.5
Internal element of compression flange	Welded b/T	≯ 23	≯ 20	≯ 18
	Rolled b/T	≯ 26	≯ 23	≯ 20
Web	Welded d/t	≯ $\dfrac{79}{0.4 + 0.6\alpha}$	≯ $\dfrac{70}{0.4 + 0.6\alpha}$	≯ $\dfrac{62}{0.4 + 0.6\alpha}$
	Rolled			

$\alpha = 2y_c/d$, where $y_c =$ distance from plastic neutral axis to edge of web connected to the compression flange instability does not occur. Members containing the amounts of plastic material necessary for the formation of plastic hinges are particularly susceptible to failure by buckling due to the large reductions in stiffness (flexural, warping, etc.) produced by the presence of these yielded regions. Thus quite severe limits on member slenderness are required if the desired form of behaviour is to be achieved. Several methods exist by which the stability of members in plastically designed structures may be ensured; these often result from rather different approaches to the problem. Prior to the publication of BS 5950 the most popular method in the UK was that of the Constrado publication *Plastic Design* [7]; this is based on the work of Horne [8–10]. Although BS 5950 does not refer directly to this, or indeed to any other methods, it effectively permits the use of any reasonable approach; the onus is then on the designer to provide justification. As an alternative to looking beyond the Code, *Cl. 5.3.5* gives an expression for the maximum distance between points of restraint L_m as

$$L_m \not> \frac{38r_y}{\left[\dfrac{f_c}{130} + \left(\dfrac{p_y}{275}\right)^2 \left(\dfrac{x}{36}\right)^2\right]^{\frac{1}{2}}} \tag{8.1}$$

in which f_c = compressive stress due to axial load (N/mm²)
$\quad\quad\;\; p_y$ = design strength (N/mm²)
$\quad\quad\;\; x$ = torsional index, see Chapter 5

For a beam ($f_c = 0$) of grade 43 steel having the fairly high value of x of 36, eqn (8.1) gives a limit of $38r_y$ which is in line with values specified in several overseas codes. Although *Cl. 5.3.5* defines restraint as 'torsional restraint', the requirement is really to prevent instability by bracing the member against both lateral deflections and twist. Bracing should always be provided in the immediate vicinity of a plastic hinge; if the actual hinge position cannot be braced then the restraint should act at a distance along the member of no more than half its depth.

Clause 5.3.6 requires web stiffeners to be provided at plastic hinges if substantial applied loads act in that immediate region, for example for a main beam supporting a single cross-beam which is transferring floor load into that main beam. Loads are regarded as substantial if they exceed 10% of the web capacity, as explained in section 5.3.1.3. The immediate region is defined as within $D/2$ of the plastic hinge point; the stiffening must also be located no further than this from the hinge point. Stiffeners should be designed as load-carrying stiffeners, as explained in section 5.3.1.2., with the additional requirement that their proportions be within the limits for plastic design sections of *Table 7*.

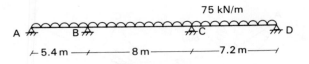

Fig. 8.5 Continuous beam of Example 8.1.

8.3 ELASTIC DESIGN OF CONTINUOUS BEAMS

Implementation of the various considerations outlined in the previous section of this chapter for the design of continuous beams on an elastic basis can most easily be appreciated by means of an illustrative example.

Example 8.1
Figure 8.5 shows a three-span continuous beam subject to a total design load of 75 kN/m over its whole length. Check whether a 457 × 191 UB 89 in Grade 43 steel would be a suitable section. It may be assumed that the beam is adequately braced against lateral deflection and twist over its whole length.

Solution
Elastic analysis using, for example, the moment distribution method [1] gives the bending moment diagram of Fig. 8.6. From this the maximum moment is 584 kN m. Since the beam is fully laterally restrained, obtain its bending strength directly from *Table 6* as $p_y = 275$ N/mm². The section under consideration is compact according to *Table 7*; therefore required section modulus is

$$S_x \not< 584 \times 10^3/275 \text{ cm}^3 = 2124 \text{ cm}^3$$

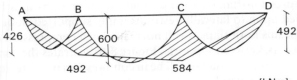

Fig. 8.6 Elastic BMD for beam of Fig. 8.5.

From section handbook for a 457 × 191 UB 89, $S_x = 2014$ cm³ and this section is therefore inadequate. A stronger section is needed such as a 456 × 191 UB 98, for which $S_x = 2232$ cm³.

Since the original section is close to being satisfactory (it is less than 5% understrength) and is compact, it is worth exploring the idea of limited redistribution of moments as permitted by *Cl. 5.4.1*. Figure 8.7 presents a modified bending moment distribution in which the support moments at B and C have been assumed to be reduced by 10% with corresponding increases in span moments.

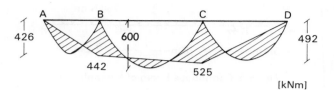

Fig. 8.7 10% redistribution of elastic BMD for beam of Fig. 8.5.

Solution
Maximum moment is now 525 kN m which requires

$$S_x \nleq 525 \times 10^3/275 \text{ cm}^3 = 1909 \text{ cm}^3$$

Taking advantage of moment redistribution therefore permits the use of a 457 × 191 UB 89.

No allowance has been made in either set of calculations for possible reductions in moment capacity due to the presence of high coexistent shear forces. This is a topic to which greater attention must be paid when designing continuous structures since the support regions will often be the critical sections. It is left to the reader to verify (using the procedure of *Cl. 4.2.3* and *Cl. 4.2.5*) that the full value of S_x may be used in this example.

The assumption that the beam of Example 8.1 was fully laterally restrained meant that design could be based on its full moment capacity. A decision on the appropriateness, or otherwise, of this assumption is often difficult for continuous beams. Indeed, because the patterns of moments will be more complex than for simply supported beams, all considerations of lateral stability will normally be less straightforward. Referring back to Chapter 5, it will be recalled that lateral instability involved both lateral deflection and twist and that an effective way of preventing its occurrence was by bracing the beam's compression flange, for example the floor slabs in a building can often be relied upon to laterally restrain simply supported floor beams. For continuous beams both flanges will normally be in compression over part of the beam's length between supports. Figure 8.6 illustrates the common situation in which the bottom flange will be in compression in the regions adjacent to the supports. A generally accepted method of assessing the susceptibility of this region to lateral–torsional instability is not presently available. For hot-rolled sections it seems reasonable to assume that the stiffness of the web will be sufficient to transfer the positional restraint provided to the top flange by the slab to the whole beam. This may not be the case for plate girders with more flexible webs. Doubtful cases should be checked using the U-frame approach of BS 5400: Part 3.

In situations where even the top flange cannot be regarded as laterally supported such as continuous crane girders, and beams not positively attached to floor systems, the beam's moment capacity must be obtained using the procedures of *Cl. 4.3*. In such cases it will often be advantageous to take account of the less severe moment diagram by modifying the effective beam slenderness using the *n*-factors of *Tables 15 and 16*.

8.4 PLASTIC DESIGN OF CONTINUOUS BEAMS

The application of plastic design to continuous beams involves two separate steps:

(1) Selection of a suitable section based on considerations of the plastic collapse mechanism.
(2) Ensuring that no other form of failure prevents the attainment of this collapse mechanism.

The first step requires the use of one of the standard techniques for plastic analysis [2–4, 7] such as the mechanism method or the reactant moment line method, while the second can largely be covered by compliance with the appropriate parts of section 5 of BS 5950.

Plastic collapse may occur in either an internal or an end span of a continuous beam. It is therefore necessary to consider the two cases illustrated in Fig. 8.8: a fixed end beam and a propped cantilever. In both cases collapse will occur when sufficient plastic hinges have formed to turn the original statically indeterminate structure into a mechanism. Figures 8.8(b) illustrate these mechanisms while Figs. 8.8(c) give the bending moment diagrams at collapse. The use of these results is illustrated by the following example.

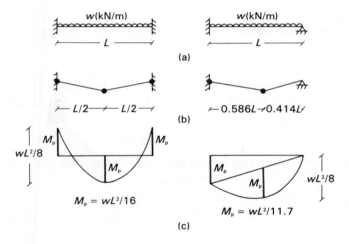

Fig. 8.8 Plastic collapse of internal and end spans of continuous beams. (a) Support arrangements; (b) plastic collapse mechanisms; (c) bending moments at collapse.

Example 8.2
Repeat Example 8.1 using plastic design.

Solution
Referring to Fig. 8.5, either span BC or span CD will govern the design.

For BC, required $M_p = wL^2/16 = 75 \times 8^2/16 = 300$ kN m

For CD, required $M_p = wL^2/11.7 = 75 \times 7.2^2/11.7 = 332$ kN m

Span CD governs, and taking $p_y = 275$ N/mm^2

Required $S_x = 332 \times 10^3/275 = 1207$ cm^3

Since S_x provided $= 2014$ cm^3, section is clearly adequate; based on considerations of moment capacity alone a 457×152 UB 60 for which $S_x = 1284$ cm^3 would be adequate. This is some 40% lighter.

The moment diagram at collapse is given as Fig. 8.9. This shows that plastic hinges would occur at C and within span CD. Without resorting to complex elastic–plastic calculations it is not possible to define precisely the remainder of the reactant moment diagram and thus the exact span moments within AB and BC. However, the moment at B must lie between the value that would just permit a plastic hinge to form within BC and M_p (for which BC would remain elastic). In either case span BC would not collapse as a third plastic hinge would be necessary to produce a mechanism.

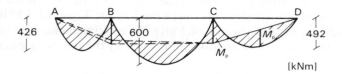

Fig. 8.9 Bending moments at collapse for beam of Fig. 8.5.

Reference to Table 8.1 shows that the revised suggestion of a $457 \times 152 \times$ UB 60 is within the cross-sectional limits for plastic hinge action. To check whether the effect of shear will reduce its moment capacity refer to *Cl. 4.2.5.*

$$P_v = 0.6 \times p_y \times D \times t$$
$$= 0.6 \times 275 \times 454.7 \times 8.0 = 600 \text{ kN}$$

No reduction in M_p is necessary if the applied shear is less than 60% of this, i.e. max. $V \not> 360$ kN. Referring to Fig. 8.9,

Shear to right of C $= 75 \times 7.2/2 + 332/7.2$
$= 316$ kN and no reduction in M_p is required.

8.5 ELASTIC DESIGN OF PORTAL FRAMES

Figure 8.10 illustrates a variety of different portal frames of the type used as the main frames in single-storey buildings. In each case the connections between the columns and the inclined rafters must be capable of transmitting bending moments if the structure is to resist horizontal loading by frame

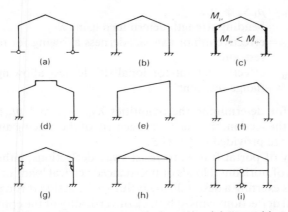

Fig. 8.10 Types of single-storey portal frames. (a) Pinned-base, (b) fixed-base, (c) heavier column sections, (d) monitor roof, (e) lean-to, (f) north light, (g) including crane, (h) tied, (i) intermediate floor. (*After ref. 12.*)

action. If such structures are to be designed elastically it will therefore be necessary to conduct a suitable analysis (or series of analyses if multiple load cases are being considered). Although traditional methods may be used, the somewhat complex geometry of the sway deflections associated with joint translation means that programs based on the matrix stiffness method are often employed nowadays.

Once the individual member forces – moments, shears and axial loads – have been determined, selection of appropriate sections proceeds very much as for members in simply designed frames. Both the columns and the rafters will normally be subject to a combination of moment and compression; they should therefore be designed as beam-columns using the procedures of Chapter 6.

When considering lateral–torsional stability, *Cl. 5.5.2* of BS 5950 permits, as an alternative to the use of the methods of Chapter 6, the use of the more favourable method of *Appendix G*. This makes some allowance for the restraining effects of the purlins, sheeting rails and cladding attached to the outer flange of the main frame members where this flange is the tension flange. For those parts of the frame where the outer flange is in compression, for example the rafters adjacent to the apex under vertical loading, it is, of course, only necessary to check stability between points of effective lateral restraint, i.e. between purlin points in most forms of construction. Thus for a uniform member, lateral–torsional buckling strength may be assessed from

$$\frac{F}{P_c} + \frac{\overline{M}}{M_b} \not> 1 \tag{8.2}$$

in which F = applied axial load
        P_c = compressive resistance determined using λ_{TC}
       $\overline{M}$ = $m_t M_{max}$

$$M_b = p_b S_x \not> p_y Z_x$$

p_b = bending strength determined using λ_{TB}

λ_{TC} = effective minor axis slenderness allowing for tension flange restraint

λ_{TB} = effective lateral–torsional slenderness allowing for tension flange restraint

Procedures for determining the quantities λ_{TC}, λ_{TB} and m_t based on the geometry of the section, the exact arrangement of the bracing and the pattern of moments are provided in *Cl. G.3*.

The ability of portal frames to resist sway deflections, either due to the direct action of horizontal loads or unbalanced vertical loads or as a result of vertical loads exerting a destabilizing influence by acting through the out-of-plumb lateral deflections caused by lack of verticality of the columns, derives principally from frame action. Although methods exist [11] for taking account of the stiffening effects of the cladding, even if such a design approach is used, a certain basic level of overall stability of the bare frames is still required. For single-bay, single-storey portals designed to a sensible limit on serviceability deflections, sway instability is unlikely to be a problem. BS 5950 does not quantify this limit for elastically designed portal frames, *Cl. 5.5.1* merely referring back to the general material on sway stability provided in *Cl. 2.4.2.3*.

Moreover, *Table 5* specifically excludes deflection limits for pitched roof portals, although it does quote $h/300$ for column-top deflections in other single-storey buildings. In the absence of specific guidance it is suggested that the limit for plastically designed frames be employed. This requires the horizontal deflection of the column tops due to a notional horizontal load equal to $\frac{1}{2}\%$ of the factored dead plus vertically imposed load to be less than $h/1000$. In calculating this deflection linear elastic analysis should be used and the stiffening effects of the cladding may be included. *Clause 5.5.3.2* gives an alternative condition in terms of an explicit expression in terms of the frame geometry and loading, the background to which is explained in reference [12]. This expression is also suitable for multibay frames for which sway instability is likely to be of greater significance [12]; such frames can also be subject to a snap-through form of instability which is covered by the provisions of *Cl. 5.5.3.3*.

8.6 PLASTIC DESIGN OF PORTAL FRAMES

Probably the most widespread application of plastic design is to single-storey portal frame structures. Often these will be the basic pitched-roof variety of Fig. 8.10(a–b). Indeed the popularity of this from of construction has been sufficient to justify the publication of several specialist texts [12, 13]. In the absence until recently of codified procedures designers have placed consider-

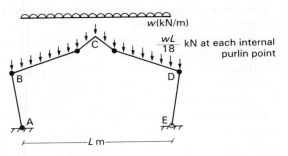

w(kN/m)

$\dfrac{wL}{18}$ kN at each internal purlin point

L m

Fig. 8.11 Typical plastic collapse mechanism for a fixed base frame under vertical load.

able reliance on the approach of the Constrado publication on plastic design [7, 14]. This in turn makes use of earlier publications on the subject by the BCSA [8, 15]. Designers wishing to use the rules of section 5 of BS 5950 to design portal frames plastically will find it necessary to refer to these earlier publications for a full treatment of the subject since the Code provides guidance on only a number of specific items. The actual application of plastic design involves the same two basic steps that were given at the start of section 8.4 on continuous beam, viz. adequate strength, avoidance of secondary failures.

For fixed-base, pitched-roof frames, collapse under vertical load will normally occur in the mode shown in Fig. 8.11. Four plastic hinges are needed to transform the frame into a mechanism (it originally had three redundancies, for example horizontal and vertical force and bending moment at the apex). Possible locations are the peaks of the elastic moment diagram, i.e. the joints, and some point within each rafter. Since vertical load is transferred from the cladding to the main frames at the purlin points, the rafter hinge(s) will form at whichever of these corresponds to the point of maximum sagging moment in the rafter. This will usually be one or two purlin points away from the apex. However, failure to locate the exact point, although it will lead to a violation of the fundamental yield condition of plastic theory [3], will normally result in only a marginal underestimate of the required M_p value. Bearing in mind the discrete nature of the available sec-

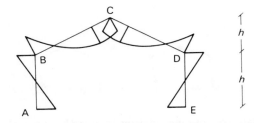

h

h

Fig. 8.12 Bending moments at collapse for frame of Fig. 8.11.

tion range, i.e. their values of S_x do not constitute a continuous spectrum, this is unlikely to prove very significant.

Application of the semi-graphical approach [12] therefore leads to the distribution of moments at collapse shown in Fig. 8.12. Knowledge of this permits the selection of a suitable section to provide the desired margin of safety. (When using a limit-states approach in which different numerical values are required for the various components of the applied loading, such as $\gamma_f = 1.4$ for dead loads and $\gamma_f = 1.6$ for imposed load, it is necessary to work with the actual factored loads; earlier publications, which used a single global load factor – typically 1.7, often showed calculations arranged in such a way that the designer could select a section to give any desired value.)

Identification of the collapse mode associated with the lowest value of the collapse load is normally quite straightforward for single-bay frames. Of the three basic modes for fixed-base frames illustrated in Fig. 8.13, mode 2 is possible only for very tall frames and/or large horizontal forces whilst mode 3 is likely only for high horizontal loading with negligible vertical load [7]. Similar behaviour is obtained for pin-base frames. Charts for the direct selection of a suitably factored value of M_p for either type of frame when subjected to a uniform vertical load plus a single horizontal eaves load are provided in reference [12]. These charts are useful for gaining an indication of the probable collapse mechanism as they permit the use of different load factors for the two types of loading. Doubtful cases can thus be identified and checked fully using a more realistic representation of wind load [7, 16]. Because of the lower load factors permitted under combined loading (see Chapter 2), most designs will be governed by the gravity load case for which good estimates of the required section, assuming a uniform frame, may be obtained from

$$M_p = \gamma_L \ \frac{wL^2}{8} \left[\frac{1}{1 + h_2/h_1 + (1 + 2h_2/h_1)^{\frac{1}{2}}} \right] \text{ for pin-base frames}$$

$$(8.3a)$$

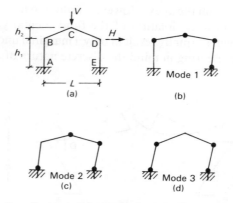

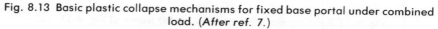

Fig. 8.13 Basic plastic collapse mechanisms for fixed base portal under combined load. (After ref. 7.)

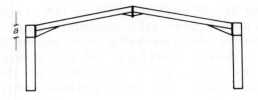

Fig. 8.14 Use of haunches in portal frame construction. (*Ref. 12.*)

$$M_p = \gamma_L \, \frac{wL^2}{8} \left[\frac{1}{1 + 0.5h_2/h_1 + (1 + h_2/h_1)^{\frac{1}{2}}} \right] \tag{8.3b}$$

in which γ_L = global load factor

and h_1 and h_2 are defined in Fig. 8.12

In deriving eqns (8.3) [15] the rafter hinges have been assumed to form at the point of maximum moment for distributed loading, i.e. the discrete nature of purlin loading has been ignored; this affects the resulting value of M_p by only a very few per cent.

For single-bay frames having other geometrices, for example those shown in Fig. 8.10, for single-bay frames having other than the same section throughout, or for multistorey frames, the reader should consult references [2, 3, 7, 12] for suitable methods of plastic analysis. These also explain the basis for the use of haunches – at either the eaves or the apex as shown in Fig. 8.14 – as a means of improving frame strength and frame stiffness. Eaves haunches are often made by splitting the basic rafter section along a diagonal and are widely used in the UK, the additional fabrication costs being more than offset by both the reduction in material and the greater rafter depth available for constructing the eaves joint. Assuming that the eaves haunches shift the eaves plastic hinges to the column top immediately below the lower end of the haunch, as shown in Fig. 8.15, for a fixed-base frame whose design is controlled by vertical load, the reduction in M_p is given by [7].

$$\frac{M_p \text{ with eaves haunch}}{M_p \text{ for uniform frame}} = \left(1 - \frac{a}{h_1} \right) \tag{8.4}$$

in which a = depth of haunch, see Fig. 8.14.

Fig. 8.15 Basic collapse mechanism (Mode 1) for haunched portal.

For typical geometries savings of between 5% and 10% on rafter size are possible. Because of the reduction in rafter moments the use of eaves haunches often leads to frames with lighter rafters than columns. Provision of a haunch at the apex does not affect the frame's basic strength (the collapse mechanism does not involve a plastic hinge at this point). However, it does reduce overall frame deflections as well as providing greater depth for the apex connection.

The types of collapse mechanism which govern the design of most portal frames are such that virtually every member is required to participate in plastic hinge action. Therefore they should each meet the cross-sectional limitations for plastic hinge action of Table 8.1. Arguments based on the identification of the last hinge to form (at which no rotation is required which, at least in theory, suggests that merely satisfying the limits for a compact section would be appropriate) should be treated with suspicion due to the difficulty of being certain that this can actually be identified [17]. Factors such as settlement, variability of material strength between members, etc., while they do not necessarily affect the plastic collapse load significantly, can change the sequence of hinge formation.

Premature failure due to lateral–torsional instability may be avoided if torsional restraints are positioned according to eqn (8.1). Purlins attached to the compression flange of a main member would normally be acceptable as providing full torsional restraint; where purlins are attached to the tension flange they should be capable of providing positional restraint to that flange but are unlikely (due to the rather light purlin/rafter connections normally employed) to be capable of preventing twist. Allowance for the limited benefit of tension flange restraint may be made by using the plastic version of the method of *Appendix G* of BS 5950. This permits torsional restraint on such members to be spaced at a distance L_t given by

$$L_t \not> \frac{L_k}{\sqrt{m_t}} \left(\frac{M_p}{M_{pr} + aF} \right)^{\frac{1}{2}} \tag{8.5}$$

in which a = distance of member axis to restraint axis (the effective position of the restraint axis may well lie beyond the member's tension flange at the line of the sheeting).

$$L_k = \frac{(5.4 + 600\, p_y/E)\, r_y x}{[5.4\,(p_y/E)\, x^2 - 1]^{\frac{1}{2}}}$$

Prior to the publication of BS 5950 the method given in reference [14], which is based on the original work of Horne [8–10], was widely used. Although not specifically mentioned by the new Code, it would appear to remain as an acceptable alternative.

Failure to meet whichever of the above criteria is selected can usually be rectified by a combination of fly braces to stabilize the main member's compression flange as shown in Fig. 8.16 and a rearrangement of the purlin spacing. The most critical area is normally the rafter adjacent to the eaves for which both modifications may be necessary. Although the methods for

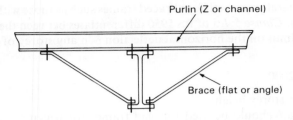

Purlin (Z or channel)

Brace (flat or angle)

Fig. 8.16 Fly-brace to compression flange of main member.

assessing lateral stability in BS 5950 are permissible (according to that document) for tapered, i.e. haunched members, experimental evidence on the performance of haunches containing plastic regions [18] suggests that instability effects are particularly severe. Until a full solution to this problem is available it is probably advisable to follow earlier advice [14, 4, 17] and ensure that haunch regions remain elastic.

8.7 ELASTIC DESIGN OF MULTISTOREY FRAMES

When designing a multistorey frame in which continuous construction is to be employed it is first necessary to establish the means by which overall stability against sway effects is to be provided. Figure 8.17 illustrates the two basic alternatives – lateral bracing in the form of concrete shear walls, a central braced core or a series of braced bays or dependence on frame action. Of these the first alternative is generally claimed to be the more economic for the types of structures built in the UK. However, client's requirements for access and utilization of internal space will sometimes be sufficiently restrictive that bracing cannot be used, in which case a sway frame will be required.

Classification of multistorey frames as 'non-sway' or 'sway' is to some extent subjective as all structures deflect laterally under the action of horizontal forces. Indeed it is possible for lightly braced frames to deform

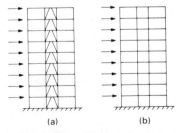

(a)   (b)

Fig. 8.17 Non-sway and sway multistorey frames: (a) non-sway, (b) sway.

more than laterally quite stiff, unbraced frames such as those with very heavy columns [19]. *Clause 5.1.3* of BS 5950 differentiates between the two classes by setting a limit on the horizontal deflection δ in any storey of a non-sway frame as

$$\delta \ngtr h/2000 \qquad\qquad\qquad (8.6)$$

in which h = storey height.

Equation 8.6 should be used for clad frames for which the calculations have been performed on the bare frame. If the stiffening effect of the cladding has been included, or the frame will not be clad in its finished condition, then the limit should be halved. Frames which do not meet this condition must be designed as sway frames. When conducting this check it is not necessary to use actual loads; notional forces equal to 0.5% of the factored dead plus imposed vertical load applied horizontally are specified in the Code. This is because the principle of the check is to assess the susceptibility of the structure to overall frame instability [20] and not to try to assess its actual sway deflections.

For non-sway frames designed on an elastic basis member forces under both vertical and horizontal loading should be determined from a linear elastic analysis of the whole frame. This is most conveniently conducted using a standard frame analysis program. In the case of regular, rectangular structures it will normally be sufficient to isolate typical frames along and across the structure and to consider planar behaviour only. Structures with more complex plan geometries or those liable to be subject to significant unbalanced loading may require at least a limited three-dimensional analysis to assess correctly the importance of overall torsional effects. *Clause 5.6.1* of BS 5950 permits two-dimensional analysis under vertical loading only to be undertaken using subframes of the type shown in Fig. 8.18 as an alternative to analysis of the full frame. In conducting the analysis under vertical loading

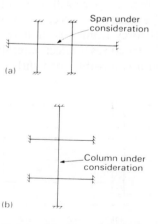

Fig. 8.18 Sub-frames for use in multi-storey frame design: (a) beam design sub-frame; (b) column design sub-frame.

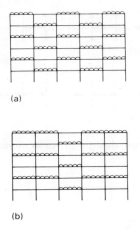

(a)

(b)

Fig. 8.19 Load patterns for beam design: (a) maximum span moments (2 such patterns required); (b) maximum support moments (6 such patterns required).

only the Code reminds the designer of the need to consider 'the most unfavourable but realistic pattern for each element', without giving details of what these patterns should be. The earlier draft [5] suggested the arrangements of Figs. 8.19 and 8.20 or if subframe analysis was being used those of Figs. 8.21 and 8.22.

Recent research [6] suggests that the identification of the particular loading arrangement that leads to the most severe combination of member forces (moment, thrust and shear) in any individual member is more difficult, the patterns of reference [5] giving underestimates in several cases. While this

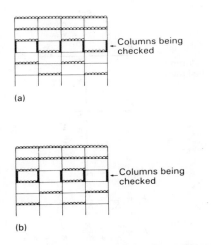

(a)

Columns being checked

(b)

Columns being checked

Fig. 8.20 Load patterns for column design: (a) single curvature bending; (b) double curvature bending.

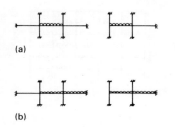

Fig. 8.21 Sub-frames for beam design: (a) span moments; (b) support moments.

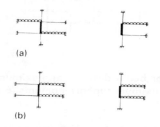

Fig. 8.22 Sub-frames for column design: (a) single curvature; (b) double curvature.

may not be of great concern for residential buildings, for which imposed load will constitute only part of the total load and extreme variations of pattern loading are unlikely, it requires attention when dealing with certain storage or industrial buildings. Up to 10% redistribution of the elastic moments is permitted providing frame members are of compact cross-section and that redistribution does not lead to reductions in column minor-axis moments. Having

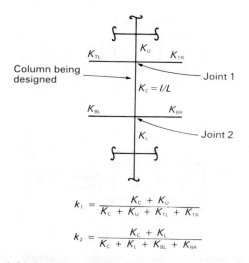

$$k_1 = \frac{K_C + K_U}{K_C + K_U + K_{TL} + K_{TR}}$$

$$k_2 = \frac{K_C + K_L}{K_C + K_L + K_{BL} + K_{BR}}$$

Fig. 8.23 Restraint coefficients for a limited frame.

determined member forces, suitable sections may be selected using the procedures already discussed in Chapters 4–6. In determining effective column
lengths the restraining effects of the beams and columns immediately
adjacent to the column being designed may be allowed for by using the
limited frame shown in Fig. 8.23 in conjunction with the effective length
chart of Fig. 8.24 [20, 21].

Sway frames should first be checked as non-sway frames using column
effective lengths from Fig. 8.24 under the most unfavourable combination(s)
of vertical load. They should then be designed for the effects of sway by considering the full vertical load to act in conjunction with the horizontal loading, including the case of notional horizontal loading without wind. The
second-order effects associated with sway deformations may approximately
be allowed for by either:

(1) Using appropriately enhanced column effective lengths, see *Appendix
E*.
(2) Amplifying the moments due to horizontal loading by a factor

$$\lambda_{cr}/(\lambda_{cr} - 1) \tag{8.7}$$

in which λ_{cr} is the load factor for elastic instability of the frame.

When using method (2), column effective lengths equal to the storey height
should be used. For the determination of λ_{cr} *Appendix F* outlines the
approximate method of Horne [12, 20, 22] which uses the sway deflections
given by a linear elastic analysis of the frame under an appropriate set of
horizontal loads to estimate the elastic critical load.

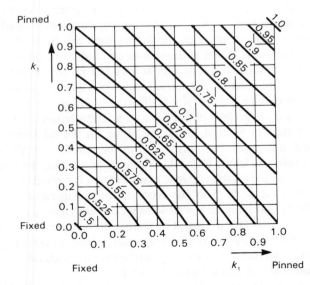

Fig. 8.24 Effective length ratio for a column in a rigid jointed non-sway frame.
(*Kirby and Nethercot, 1979.*)

8.8   PLASTIC DESIGN OF MULTISTOREY FRAMES

Provided the structure satisfies the requirement of eqn (8.8) for treatment as a non-sway frame it may be designed plastically without specific consideration of its response to lateral loading. Determination of the plastic collapse load will require the application of one of the standard plastic analysis techniques [2, 3, 7, 12]. Because the presence of significant amounts of plasticity in the columns can cause large reductions in stability unless the design is carefully controlled, it is normal practice to limit plastic hinge action to the beams. In such cases the application of plastic design effectively consists of treating the beams as continuous over the columns and designing for vertical load only. Two methods of doing this for an eight-storey frame example are presented in reference [7]. An alternative approach which makes use of continuity in both planes of the (assumed rectangular in plan) frame is the subject of a joint Institution of Structural Engineers and Institute of Welding report [23]. As a result of two series of full-scale tests [24, 25], improvements to the method of column design in that report have recently been proposed [26].

Plastic design of sway frames is a complex topic which is still the subject of much research. The central problem is the need to make suitable allowance for the effects of sway. Therefore before attempting to design on this basis it is necessary to acquire a proper understanding of the structural actions involved, for example by studying the relevant sections of reference [3, 12, 13, 20, 21]. Simply attempting to use the material of *section 5.7.3* of BS 5950 without this basis is likely to lead to misapplication of the rules.

The method of BS 5950 uses the empirically based Merchant–Rankine formula [12, 20, 22] to assess the severity of sway effects. This formula is based on the premise that since collapse occurs by an interaction of plasticity and instability a good estimate of the failure load can be obtained from a knowledge of the simple plastic collapse load and the elastic critical load. The former gives the collapse load for very stocky frames (for which instability effects are negligible) while the latter provides a good indication of the strength of very slender frames (which effectively fail by elastic instability). Thus, providing λ_{cr} is greater than 10 for an analysis based on a bare frame that will subsequently be clad, or 20 for either an unclad frame or a clad frame in which allowance has been made in the analysis for the stiffening effect of the cladding, frame instability has negligible effect and design may be based on the simple plastic collapse load. Frames for which λ_{cr} is less than 4.6 in the first case or 5.75 in the second would experience such severe instability effects that their design must be based on a rigorous second-order elastic–plastic analysis [12, 27, 28]. For frames in the intermediate range the full analysis is not required providing the collapse load is taken as the simple plastic collapse load multiplied by a reduction factor which takes account of the limited influence of frame instability [20].

REFERENCES

1. Coates, R.C., Coutie, M.G. and Kong, F.K. *Structural Analysis* 2nd edn. Van Nostrand Reinhold (UK), Wokingham (1980).
2. Neal, B.G. *The Plastic Methods of Structural Analysis*. Chapman & Hall, London (1963).
3. Horne, M.R. *Plastic Theory of Structures*. Pergamon Press, Oxford (1978).
4. Morris, L.J. A commentary on portal frame design. *The Structural Engineer* **59A** (12), 394–403 (December 1981).
5. B/20, *Draft standard specification for the structural use of steelwork in building: Part 1: Simple construction and continuous construction*. BSI, London (1978).
6. Yau, F., Hart, D.A., Kirby, P.A. and Nethercot, D.A. Influence of loading patterns on column design in multi-storey rigid-jointed steel frames. In *Instability and Plastic Collapse of Structures* edited by Morris, L.J. Granada, London (1983) 232–42.
7. Morris, L.J. and Randall, A.L. *Plastic Design*. Constrado Publication (1979).
8. Horne, M.R. *The Plastic Design of Columns*. BCSA Publication No. 23, (1964).
9. Horne, M.R. The stanchion problem in frame structures designed according to ultimate carrying capacity. *Proc. Instn. Civil Eng.* 5 (1), Part III, 105–60 (April 1956).
10. Horne, M.R. Safe loads on I-section columns in structures designed by plastic theory. *Proc. Instn. Civil Eng.*, 137–50 (September 1964).
11. Bryan, E.R. and Davies, J.M. *Manual of Stressed Skin Diaphragm Design*. Granada Publishing, London (1982).
12. Horne, M.R. and Morris. L.J. *Plastic Design of Low Rise Frames*. Granada, London (1981).
13. Heyman, J. *Plastic Design of Structures*, vols. 1 and 2, Cambridge University Press, Cambridge (1969).
14. Morris, L.J. and Randal, A.L. *Plastic Design*. Constrado, London, (1979), (Supplement).
15. Horne, M.R. and Chin, M.W. *Plastic Design of Portal Frames in Steel to BS 968*. BCSA Publication No. 29 (1966).
16. CP3: Chapter V, *British Standard Code of Basic Data for the Design of Buildings: Part 2: 1972 Wind Loads*. BSI, London (1972).
17. Morris, L.J. A commentary on portal frame design. Discussion, *The Structural Engineer* **61A** (6, 7), 181–9, 212–21 (June, July 1983).
18. Morris, L.J. and Nakane, K. Experimental behaviour of haunched members. In *Instability and Plastic Collapse of Structures* edited by Morris, L.J. Granada, London 547–59 (1983).
19. Massonnet, C. European recommendations for the plastic design of steel frames. *Acier-Stahl-Steel* (4), 146–53 (1976).
20. Kirby, P.A. and Nethercot, D.A. *Design for Structural Stability*. Granada, London (1979).
21. Wood, R.H. Effective lengths of columns in multi-storey buildings. *The Structural Engineer* **52**, (7, 8, 9), 235–43, 295–302, 341–6, (July, August, September 1974).
22. Horne, M.R. An approximate method for calculating the elastic critical loads of multi-storey plane frames. *The Structural Engineer* **53**, 242–8 (1975).
23. The Institution of Structural Engineers and The Institute of Welding. 2nd Joint Report on fully rigid multi-storey welded steel frames (May 1971).
24. Wood, R.H., Needham, F.H. and Smith, R.F. Test of a multi-storey rigid steel frame. *The Structural Engineer* **46** (6), 107–19 (June 1968).
25. Smith, R.F. and Roberts, E.H. Test of a fully continuous multi-storey frame of

high yield steel. *The Structural Engineer* **49** (10), 451–66 (October 1971).
26. Wood, R.H. *A New Approach to Column Design.* HMSO, London (1973).
27. Vogel, U. Recent ECCS developments for simplified second-order elastic and elastic–plastic analysis of sway frames. *Third Int. Coll. Stability of Metal Structures*, Paris, 217–24 (1983).
28. Majid, K.I. and Anderson, D. The computer analysis of large multi-storey framed structures. *The Structural Engineer* **46**, 357–69 (1968).

EXAMPLES FOR PRACTICE

1. Determine the maximum spacing between points of effective lateral restraint for a 305 × 102 UB 25 in grade 43 steel if such a section is used as a beam required to participate in plastic hinge action.

[0.60 m]

2. Check whether a 254 × 254 UC 89 in grade 50 steel carrying an axial load of 1400 kN over a free height of 2.5 m is capable of participating in plastic hinge action.

[No, L_m limit is 2.23 m]

3. Select a suitable UB in grade 43 steel to act as a 3-span continuous beam having spans of 7.2 m, 10.4 m and 6.5 m, assuming loading over the whole beam of 57 kN/m based on (i) elastic design, (ii) plastic design. In both cases you may assume full lateral restraint to all spans.

[457 × 191 UB 82 or 457 × 191 UB 74 if redistribution is used,
457 × 191 UB 67]

Index